Fourier Transform
NMR Techniques:
A Practical Approach

Fourier Transform NMR Techniques: A Practical Approach

K. MÜLLEN
P. S. PREGOSIN

Laboratorium für Anorganische und Organische Chemie,
Eidg. Technische Hochschule, Zurich,
Switzerland

1976

ACADEMIC PRESS
LONDON · NEW YORK · SAN FRANCISCO
A Subsidiary of Harcourt Brace Jovanovich, Publishers

ACADEMIC PRESS INC. (LONDON) LTD.
24–28 Oval Road,
London NW1

United States edition published by
ACADEMIC PRESS INC.
111 Fifth Avenue
New York, New York 10003

Library of Congress Catalog Card Number 76 20280
ISBN: 0 12 510450 2

Printed in Great Britain by
C. F. Hodgson & Son Ltd., 50 Holloway Road, N7 8JL, England

Preface

" Think small " suggests one manufacturer of nuclear magnetic resonance (nmr) spectrometers, as he advertises that his instrument can measure proton nmr spectra on samples containing 50 μg or less. No less routine are commercial inserts suggesting that ^{13}C nmr may provide the solution to ones research problem. These suggestions only serve to highlight that the face of nmr has recently undergone an uplift. The current literature now shows that where, previously, only relatively few nuclei were observed directly, it is now possible readily to observe some 10–20 different species. These gains, for the chemist, are the direct result of what has come to be known as Fourier transform nmr.

In retrospect, we clearly see that many of the concepts involved in Fourier transform nmr have been employed extensively in other areas for some years. The interface of these and newer techniques has, however, had a major effect on the shape of the modern nmr spectrometer. Despite these and the con-tinuing changes in the " state of the art ", a few fundamental principles are all that is necessary to operate such a spectrometer. We feel that these concepts and the actual mechanics of operation are well within the grasp of any chemist. Although the myriad of " blinking lights " may create the impression that a thorough grounding in both quantum mechanics and electrical engineering are prerequisites for measuring a spectrum, this is not true. Indeed we have found that best results are frequently forth-coming from an operator whose chemical background is strong, since it is he who best knows what phenomena are peculiar to his research area.

With the chemist in mind we have attempted, in this text, to set down, qualitatively, these few basic principles. In addition to these we have included a short practical introduction to both the computer and the spectrometer based on our experiences with Varian XL-100 and Bruker HX-90 type spectrometers, as well as a section on the newer benefits which can accrue for the chemist after his mastery of the technique. In a short text it is obviously difficult to provide great detail in any one subject, and for in-depth coverage in any one area we recommend that the reader refer to the literature cited at the end of each chapter.

Put simply we hope, here, to provide the chemist with a practical introduction to the technique after which he will be able to move, in only a few jumps, directly to the spectrometer.

We should like to thank Dr. E. Becker, Prof. R. R. Ernst, Dr. H. Kellerhals, Prof. G. Lev, Prof. J. F. M. Oth, Prof. E. W. Randall, Prof. L. M. Venanzi and Dr. F. Wehrli for helpful discussions during the preparation of the manuscript. Additional thanks are due to Mr. G. Balimann for both his useful commentary and excellent artwork and Mrs. B. Jägerhofer for typing the manuscript.

September 1976   K. MÜLLEN and P. S. PREGOSIN

Foreword

Perhaps no one has been more surprised and delighted at the continued virility, interest and utility of the nuclear magnetic resonance technique than the practitioner himself. He has been favoured with renewed interest and fascination in his subject like the fortunate husband captivated by the undiminished, in fact increased, vivacity of his wife. The large and continued number of new developments both in technique and application have meant that the useful lifetime of many a good nmr text has shortened. This has been a partial justification for the number of publications on this topic, but it is certainly not a complete one, and the informed reader will look for some originality in any new text, either of presentation and coverage or of insight.

Dr. Müllen and Dr. Pregosin have identified a real gap in the coverage of the topic between the practical (at the level of the manufacturers instruction manual) and the applied (as represented by the various reviews of the chemical application of the technique). They do fill, in my view, this gap with sufficient insights to justify another addition to the nmr literature. Their volume therefore should be of considerable use to the chemist, newly graduated or recently converted, who begins to address himself to the question of how to obtain meaningful nmr results.

September 1976   E. W. RANDALL

To our wives, whose patience and encouragement have made this text possible.

Contents

Chapter 1 CW and pulsed nmr. A general description

Chapter 2 The role of the computer

Chapter 3 Spectrometer operation

Chapter 4 Spin lattice relaxation times

Chapter 5 Commonly occurring non-routine problems

Chapter 6 "New nuclei". A bonus in the solution of chemical problems

Chapter 7 Dynamics and reaction mechanisms

1
CW and pulsed nmr.
A general description

1.1 Introduction

Most forms of spectroscopy make use of the fundamental principle that molecular systems can only exist in discrete energy states. This is advantageous since transitions between these states can be induced by providing the system with the exact energy of this separation. Changes in the macroscopic properties which occur at this time can be recorded experimentally, perhaps as a function of the excitation energy, and consequently interpreted in terms of molecular and electronic structure.

Such a property is the magnetism of atomic nuclei in a static magnetic field H_0.[1] The interaction of a nuclear magnetic moment, μ, of spin $I = \frac{1}{2}$ with the field H_0 leads to two states of different energy. In an ensemble of identical nuclei the population of these states, according to the Boltzmann-law, depends on their energy separation, ΔE as shown in eqn (1.1)

$$\Delta E = 2\mu H_0. \tag{1.1}$$

Substituting for μ,

$$\mu = \gamma h I/2, \tag{1.2}$$

where

$$\gamma = \text{the gyromagnetic ratio of the nucleus}$$

$$I = \text{the nuclear spin } (= \tfrac{1}{2}),$$

affords

$$\Delta E = \gamma h H_0/2\pi. \tag{1.3}$$

Since

$$\Delta E = h\nu, \tag{1.4}$$

we come to the well known condition

$$\nu = -\gamma H_0/2\pi \tag{1.5}$$

or

$$\omega_0 \text{ (angular frequency)} = -\gamma H_0 \tag{1.6}$$

relating the applied field, H_0, to the frequency of resonance.† The representation of $-\omega/\gamma$ as a field strength will prove useful later.

On the average a single nucleus remains in a certain state no longer than a time, T_1, the spin lattice relaxation time. The interactions of the spins with their surroundings, the lattice, represent the processes by which they exchange energy to reach equilibrium. Therefore, as is shown in eqns (1.7) and (1.8), the nmr line width will be proportional to $1/T_1$. Since

$$\Delta t \Delta E \geqslant h \text{ (uncertainty principle)}, \qquad (1.7)$$

if we set $T_1 = \Delta t$, then

$$\left. \begin{array}{c} T_1 \, \Delta E \geqslant h \\[6pt] T_1(h \Delta v) \geqslant h \\[6pt] \Delta v \geqslant 1/T_1. \end{array} \right\} \qquad (1.8)$$

Normally the resonances are significantly broader than $1/T_1$ due to interactions between the different nuclear spins. These processes which allow the spins to come to equilibrium with each other are characterized by a second time constant, T_2, the spin–spin relaxation time.

If there were no selective screening of H_0 in eqn (1.6), all protons (or carbons, or nitrogens ...) would resonate at the same frequency. However, there arises at each nucleus a local field due to the surrounding electrons, and due to other nuclei present in the molecule, which modifies the external field H_0. Thus, depending upon the differential shielding by the electrons and on the type of spin–spin coupling with other nuclei, the spins absorb at different frequencies giving rise to the well known nmr-spectrum. In addition, if we recall that the absorption of energy at a given frequency is proportional to the number of nuclei in a given magnetic site, we have an elementary picture of the three most important nmr spectral parameters: the chemical shift, the coupling constant and the signal intensity.

1.2 The time dependence of the nuclear magnetization

We must now modify our mental picture to include the fact that the nuclear magnets are in motion. An ordinary bar magnet exposed to a static magnetic field H_0 will tend to align with it. However, if the magnet possesses an angular momentum, L, as does the nuclear magnet, it will act like a gyroscope in a gravitational field and instead of being aligned, it will precede around the direction of H_0 with a characteristic frequency ω.

The equation for this motion can be derived simply by considering that

† The insertion of the negative sign is meant to indicate clockwise rotation.

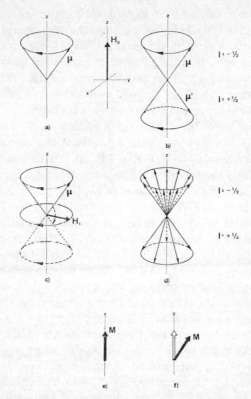

FIG. 1.1. The motion of the nuclear spins in the presence of the stationary field, H_0, and the transmitter field, H_1. (a) A single nuclear magnet in the presence of H_0. (b) Two nuclear magnets; one in each of the two spin states. (c) One nuclear magnet in the presence of H_0 and H_1. At resonance (dotted) this is tipped away from its original position. (d) An ensemble of nuclear magnets. (e) Vector expressing net nuclear magnetization. (f) Vector representing net nuclear magnetization at resonance.

the torque, $\mu \times H_0$ exerted on μ by the field equals the rate of change of angular momentum

$$d\boldsymbol{L}/dt = \boldsymbol{\mu} \times \boldsymbol{H}_0.\qquad(1.9)$$

Since

$$\boldsymbol{\mu} = \gamma.\boldsymbol{L}\qquad(1.10)$$

then

$$d\boldsymbol{\mu}/dt = \boldsymbol{\mu} \times (\gamma \boldsymbol{H}_0).\qquad(1.11)$$

The relationship (1.11) is valid not only for the case where the field is time independent, but also for the general case when the field is composed of more than one component, one of which is time dependent. The motion for such a single magnetic moment is shown in Fig. 1.1a.

Naturally, since μ may also be aligned against the field, a more accurate picture is Fig. 1.1b in which we portray the two energetically different possibilities for our magnet in the external field. If we now apply an alternating transmitter field, H_1 rotating at ω, the so-called Larmor frequency of the nucleus, (Fig. 1.1c) μ will see two apparently static fields (H_1 is moving with μ) and the magnet will be "tipped" away from its equilibrium position. In this fashion the energy of our small magnet is changed.

Realistically, we should first consider the effect of H_0 on an *ensemble* of nuclei (Fig. 1.1d). For a nucleus with a positive γ the excess of spins resides in the state having spin $I = -\frac{1}{2}$(Fig. 1.1d). In this case the magnetization, M, which represents the vector sum of all the magnetic moments, takes the form of a vector aligned in the direction of H_0 (Fig. 1.1e). No net magnetization is observed in the XY plane.

As a consequence of the perturbation by the alternating field H_1, the magnetization vector M is tipped away from the direction of H_0 producing components in the xy plane (Fig. 1.1f). Immediately following this process the transverse components M_x and M_y decay to zero with a time constant T_2, while the longitudinal component, M_z, is restored to its equilibrium value with a time constant, T_1. After some time sufficiently greater than T_1, the equilibrium situation will have been restored.

To mathematically present the time dependence of the motion of the magnetization vector, M, requires that we modify our eqn (1.11) to include both a new total field, H, one part of which is time dependent, and the rate processes of relaxation characterized by T_1 and T_2. This has been done;[2] however, the results are perhaps best appreciated by considering them in a rotating frame of reference.[3, 4] By this we mean considering these phenomena, not in a fixed rectangular coordinate system X, Y, Z but rather in one which rotates around the z axis (as does our H_1 field), with frequency ω and with the new coordinates x, y, z.

Within this rotating frame of reference it can be shown, considering the total magnetization M and not just one magnetic moment, that the differential eqn (1.11) becomes eqn (1.12),

$$\frac{\mathrm{d}M}{\mathrm{d}t} = \gamma M \times H_{\mathrm{eff}} \qquad (1.12)$$

with

$$H_{\mathrm{eff}} = H_0 + H_1 + \omega/\gamma \qquad (1.13)$$

where the term ω/γ, the so-called fictitious field, is introduced as a consequence of the change in coordinate system. Such a manipulation in no way affects the physical reality of the experiment. We are only positioning ourselves in such a way that we may more clearly "watch" the motion of the magnetization vector without the complication of the time dependence of the transmitter field, H_1.

For the special case where $\omega = \omega_0$, the fictitious field will cancel H_0 ($= -\omega_0/\gamma$) and eqn (1.13) reduces to $H_{eff} = H_1$. In the rotating frame M will then precede around H_1, the applied field, whose direction we shall arbitrarily assign to the x axis. The angle, α, through which M is tipped, and consequently the magnitude of the magnetization vector in the y direction (our signal!), depends upon the strength as well as the length of time, t_p, during which H_1 is applied according to eqn (1.14).

$$\alpha = \gamma H_1 t_p. \tag{1.14}$$

The rotating frame concept can be quite useful in describing the behaviour of the magnetization vector when the method of applying H_1 is varied.

1.3 H_1 continuous

For the case where ω/γ cancels H_0 the magnetization vector is tipped into the y direction and, given a suitably placed receiver coil, we detect a signal. For H_0 far from resonance H_{eff} is essentially H_0 and the vector M is oriented in the z direction. However, if we "sweep" ω_0/γ such that we slowly approach resonance, H_{eff} begins to deviate from the z direction and with it, the magnetization. Where the field ω_0/γ changes slowly enough such that the magnetization vector can follow H_{eff}, we have the so-called adiabatic or slow passage experiment,† during which time H_1 is constantly "on".[3] The sweep is continued until all of the desired nuclei have been successively excited. Since the alteration in H_0 may be accomplished either by changing H_0 (field sweep) or ω_0 (frequency sweep), we have two methods for performing this experiment. It is essential that the transmitter and detector work monochromatically, such that any occurring transition is separately excited and registered.

One fundamental disadvantage of such a continuous wave (CW) experiment is the time demand, commonly of the order of 250–500 s/sweep.[5] This follows naturally from the necessity of irradiating each chemically shifted nuclear spin separately. Additionally we know that the rate of absorption of energy by a spin system depends both on T_1 and the transition rate, the probability, W, per unit time of a given transition.[4] We know also that W is proportional to the square of the exciting field, H_1, and we might therefore expect that the absorption of energy by the system, and thus the nmr signal, might be increased by increasing H_1. Unfortunately, when W is of the order of $1/2T_1$,[4] the T_1 process no longer suffices to maintain the population

† In reality significantly faster sweep rates than required by this condition ($\gamma H_1^2 \gg dH_0/dt$) are normally employed.

excess in the lower level and the system becomes "saturated". Such a limitation can seriously affect the sensitivity† of the experiment.[1]

1.4 H_1 pulsed‡

A novel conceptual approach to the question of sensitivity and the nmr experiment is based on the so-called Felgett-principle.[6] If it is possible to simultaneously excite the entire spin ensemble with n transmitters, the S/N ratio obtainable is increased in proportion to $\sqrt{n}$. For each of the n transmitters we should have a single receiver which is tuned to receive only information coming from the corresponding transmitter. The impracticability of such an arrangement is obvious. On the other hand we can simulate the n transmitters via one high power radiofrequency pulse. With this we enter the realm of high resolution pulsed nmr spectroscopy.[4] In a pulsed nmr experiment a strong transmitter field H_1 is applied for a short time t_p. This will, as before, rotate the magnetization vector by an angle $\alpha = H_1 t_p$ away from the equilibrium orientation. The transverse magnetization that is generated has its maximum value when the product of t_p and γH_1 is such that $\alpha = 90°$. This we commonly call a 90°-pulse. The effect, on a typical spectrum, of varying α is shown in Fig. 1.2.

If $\alpha = 180°$, the pulse would simply invert the magnetization without generating a signal. Quite different from the condition of an adiabatic passage, this strong H_1-field (typical field strengths in the pulsed experiment are of the order of several G, whereas in the CW experiments, H_1 is of the order of mG) is "on" only for a few microseconds. Since H_{eff} is very rapidly changing from the z toward the x direction, the magnetization vector cannot follow H_{eff} but precedes around the vector of H_{eff} that is left at any time after the removal of H_1. Since the transverse magnetization after the pulse decays exponentially to zero, due to spin-spin relaxation, the receiver coil will now register the output signal as a "free induction decay". The decay is "free" in that the spins precede in the absence of an RF-field. It is this decay, although it is a time and not a frequency domain function, which contains the spectral information we are seeking.

The condition for an "ideal" pulse which is able to rotate all spins by the

---

† Sensitivity is the ability of the spectrometer to differentiate the signal from the surrounding noise. It is usually defined as the ratio (signal intensity/root mean square (rms) noise), where the approximation for rms noise is rms noise ≈ peak-to-peak noise/2·5. One concludes that sensitivity may be increased either by increasing the signal intensity or decreasing the noise.

‡ For a more thorough presentation concerned with pulsed nmr spectroscopy the reader is recommended to consult references 3 and 7–12.

same angle is

$$\gamma H_1 \gg 2\pi\Delta,$$

whereby Δ is the range of Larmor frequencies.

When the transmitter frequency is not identical to the Larmor frequency we obtain an interference of M_y and H_1 and record a sine wave with an exponentially decaying amplitude.[3, 7, 12]. As may be seen from Fig. 1.3, the distance between the higher frequency maxima is exactly the reciprocal value

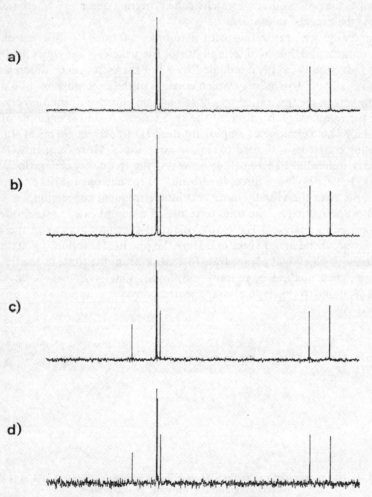

a)

b)

c)

d)

FIG. 1.2. The effect of α, the " flip "-angle, on the ^{13}C frequency spectrum of ethyl benzene (one pulse spectra). $\alpha =$ (a) 90°, (b) 60°, (c) 30°, (d) 10°. Spectra are scaled to a single intensity.

of the difference between the H_1 and Larmor frequencies. The function in Fig. 1.3(a) may be analytically described as†

$$M_y(t) = M_{y_0} \cdot [\cos{(\omega_0 - \omega_1)t}] \cdot e^{-t/T_2^*}, \qquad (1.15)$$

in which M_{y_0} is the transverse magnetization immediately after the pulse and is equal to

$$M_{y_0} = M_{z_0} \cdot \cos{\alpha} = M_{z_0} \cdot \cos{\gamma H_1 t_p} \qquad (1.16)$$

and M_{z_0} is the equilibrium magnetization in the z direction. Equation (1.15) describes the process that we watch as the "rotating observer" where $(\omega_0 - \omega_1)$ is now the precession frequency.

Generally, we are dealing with a number of spins of different chemical environment and thus of differing Larmor frequencies. Consequently, the FID (free induction decay) is modulated by all of these frequency differences and we normally observe a decay which consists of the superposition of a number of sine waves, the separation of whose maxima affords information concerned with the frequency separation of the resonances. A simple example of this is Fig. 1.3. The technique of employing the FID to extract chemical shifts and coupling constants is limited to simple molecules. More frequently the FID appears somewhat like Fig. 1.4; however, the frequency separations shown in Fig. 1.3 are no less a direct "read-out" of the chemical shifts and coupling constants than the visually more acceptable frequency spectrum.

To properly analyse the time response of a complicated system one needs a broad band analyser. The employment of specific filters as in the optical field is unrealistic in our situation. Here the practical solution is a mathematical analysis which will permit transformation from the time, to the frequency domain. The method commonly employed, that is Fourier analysis, has given its name to this type of nmr spectroscopy.[5]

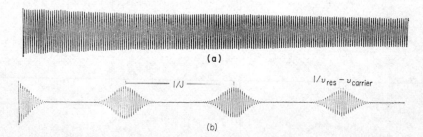

FIG. 1.3. The appearance of the $P(OCH_3)_3$ ^{31}P free induction decay (FID). (a) With ^1H decoupling and (b) without ^1H decoupling.

† In the expression for the exponential decay of the transverse magnetization a time constant T_2^* instead of T_2 is commonly applied. This includes the effects of spin-spin relaxation and magnetic field inhomogeneity.

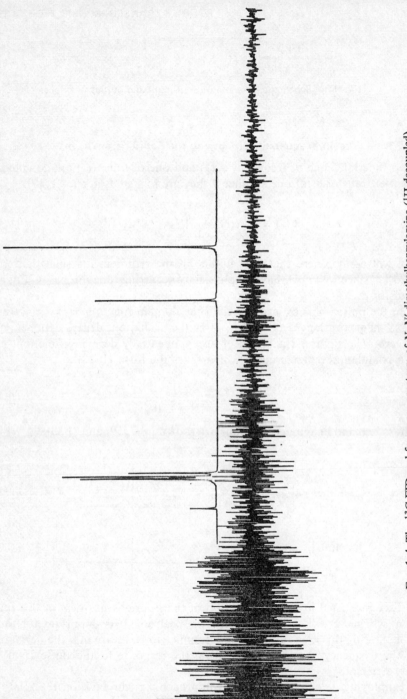

FIG. 1.4. The ^{13}C FID and frequency spectrum of N,N-dimethylbenzylamine (^{1}H decoupled)

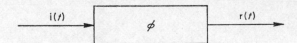

FIG. 1.5. Schematic representation of input and output functions.

1.5 Time domain, frequency domain and the Fourier transform

Two functions, one of frequency $X(v)$ and one of time $x(t)$ are accepted as Fourier transforms of one another if they are related as in eqn (1.17).[13]

$$X(v) = F\{x(t)\} = \int\limits_{-\infty}^{+\infty} e^{-2\pi i v t} x(t)\, dt. \qquad (1.17)$$

We will assume here that $x(t)$ fulfills all the requirements such that $X(v)$ "exists" since we know that any physically occurring function has its Fourier transform.

In the pulsed nmr experiment the time domain function $x(t)$ is the transverse magnetization $M_y(t)$ which, after the excitation, decays exponentially to zero. The Fourier transform of an exponentially decaying function, e^{-t}, is a complex Lorentzian curve which takes the form (1.18)

$$F\{e^{-t}\} = \frac{1}{1+2\pi i v} \qquad (1.18)$$

whose real and imaginary solutions are shown in (1.19) and (1.20).

$$\text{Real} \propto \frac{1}{1+(2\pi v)^2} \qquad (1.19)$$

$$\text{Imag.} \propto -\frac{2\pi i v}{1+(2\pi v)^2} \qquad (1.20)$$

We are interested in the real solution of the Fourier transform of this function, which we will call the absorption spectrum. It is important for those first entering this area of magnetic resonance to be aware that the derivation of the frequency spectrum, $X(v)$, from the response to the pulse is in no way artificial.

Employing a somewhat different approach we can envision the relationship between the frequency and time spectra by considering the question of how a linear† system, represented in a first approximation by an ensemble of spins, can be characterized.[1a, 10, 14 a, b]

The "response" function $r(t)$ of such a system, that is the "reply" of a system to excitation by an "input" function $i(t)$, can be mathematically formulated by application of a system-specific operator ϕ on $i(t)$.

If we choose as the input function, $i(t)$, an eigenfunction of the system then the resulting "response" takes the form of eqn (1.21);

$$\phi i(t) = S\, i(t) \tag{1.21}$$

where S is some constant of interest. In the CW nmr experiment where the eigenfunction is of the form $e^{2\pi i v t}$, the response of the system, $r(t)$, appears as:

$$r(t) = S(2\pi i v)e^{2\pi i v t}, \tag{1.22}$$

where S is now the complex frequency spectrum containing an absorption and a dispersion component.[15] If we choose a complete set of basis functions as input, instead of a single expression, then we obtain as the response:

$$r(t) = \int_{-\infty}^{+\infty} S(2\pi i v)\, e^{2\pi i v t}\, dv \tag{1.23}$$

where $S(2\pi i v)$ is the frequency-response.

While the input function in the pulsed experiment is not an eigenfunction of our system it has been shown that the output function, r, appears as an equation similar to that of (1.22) except that the frequency-response $S(v)$ is now replaced by a pulse response $s(t)$ which is not a function of frequency but rather one of time.

If indeed the two types of responses yield identical information with the exception of their form it should be possible to find a simple expression which connects these two. This is shown in eqn (1.24),[16]

$$S(v) = \int_{-\infty}^{+\infty} e^{-2\pi i v t} s(t)\, dt, \tag{1.24}$$

and is readily recognized to be eqn (1.17) "dressed" slightly differently.

The primary virtues of the pulse technique lie:

(a) In the speed with which data may be taken (generally no more than several seconds as opposed to several hundreds of seconds in the CW-mode).

† In a very elementary way one can think of a linear system as one for which the output is directly proportional to the input.

(b) In the ability to excite all occuring transitions simultaneously, whether they encompass 500 or 15 000 Hz spectral widths.

Since the decay of the transverse magnetization after a pulse, and thus the acquisition of the spectral information, takes much less time than does a CW-sweep, pulsed FT spectroscopy is especially suitable for spectrum averaging. By "averaging" we refer to the coherent summation of a successive series of spectra in order to improve the S/N ratio in the final frequency spectrum. In a fundamental work involving FT nmr,[5] it has been shown that for a given sensitivity this method affords a time gain of the order of magnitude $\Delta/v_{\frac{1}{2}}$, where Δ is the chemical shift range to be measured and $v_{\frac{1}{2}}$ is the width at half height of the spectral lines. Thus for protons, where $v_{\frac{1}{2}} \sim 1$ Hz and $\Delta \sim 1000$ Hz, this may represent a time saving of the order of 10^3. If one allows that the time slow sweep conditions in CW nmr are not usually met this gain is reduced by approximately one order of magnitude to 10^2.[9]

Table 1.1

Values for $\sin \alpha$ and $(1 - \cos \alpha)$

α	$\sin \alpha$	$(1 - \cos \alpha)$
5	0·087	0·004
10	0·174	0·015
15	0·259	0·034
20	0·342	0·060
30	0·500	0·134

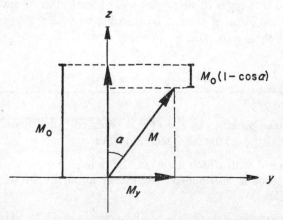

FIG. 1.6. The components of magnetization, M_y and M_z.

Obviously the time gain, and thus the difference between being able to undertake a detailed nitrogen or carbon nmr study and perhaps making only an occasional measurement, is greatest for nuclei with wide ranges of chemical shift. The effects of saturation, touched on in Section 1.3, are also observed in the FT nmr experiment.[17] In this mode coupled spin systems saturate homogeneously so that no change in the relative intensities of the signals is found; however, in the CW mode the effects are non uniform.

The implication of signal averaging in the FT mode is that we will be subjecting the spin system to a repetitive sequence of pulses.[5, 1a, 18b] If one uses 90° pulses and is careless about the selection of the interval between the pulses (e.g. T_1 processes are not yet complete although the signal in the y direction has decayed to zero) the signal which results in the y direction after each pulse will be attenuated. Since the main goal of signal averaging is the increase of S/N, we must compromise. It has been shown that under the conditions, $\alpha = 90°$, and the pulse interval $< T_1$, a steady state is built up which will result in a signal intensity proportional to (pulse interval†/$2T_1$).[19]

Choosing $\alpha < 90°$; will, after one pulse, attenuate the transverse magnetization M_y by a factor $\sin \alpha$ (see Fig. 1.6) while at the same time diminishing M_z by $M_{z_0} (1 - \cos \alpha)$. If we now choose α such that $\sin \alpha \gg (1 - \cos \alpha)$ (i.e. α is small), the loss in sensitivity due to the decrease in M_y is compensated by the gain in M_z due to the spin-lattice relaxation after the first pulse, but before the second pulse.[3] Table 1.1 illustrates the situation for various values of α. Thus in the multipulse experiment, we are deliberately reducing the potential magnitude of M_y to avoid waiting times $> T_1$ for the system to equilibrate. Some practical considerations for dealing with the choice of the flip angle, α, and the timing of pulse and the acquisition of data will be provided in Chapter 3.

References

1. The reader will find the following texts useful for reviewing fundamental concepts:
1a. A. Abragam, "The Principles of Nuclear Magnetism." Clarendon Press, Oxford, 1961.
1b. J. A. Pople, W. G. Schneider and H. J. Bernstein, "High Resolution Nuclear Magnetic Resonance." McGraw-Hill, New York, 1959.
1c. J. W. Emsley, J. Feeney and L. H. Sutcliffe, "High Resolution Nuclear Magnetic Resonance Spectroscopy," Vols. 1 and 2. Pergamon Press, Oxford, 1965.
1d. F. A. Bovey, "Nuclear Magnetic Resonance Spectroscopy." Academic Press, New York and London, 1969.
1e. E. D. Becker, "High Resolution NMR." Academic Press, New York and London, 1969.
1f. H. Guenther, "NMR-Spektroskopie." G. Thieme Verlag, Stuttgart, 1973.

† Here we take the pulse interval to be the acquisition time, t_{acq}.

2. F. Bloch, *Phys. Rev.*, **70,** 460 (1946); F. Bloch, W. W. Hansen and M. Packard, ibid., **70,** 474 (1946).

3. T. C. Farrar and E. D. Becker, "Pulse and Fourier Transform NMR." Academic Press, New York and London, 1971.

4. C. P. Slichter, "Principles of Magnetic Resonance." Harper and Row, New York, 1963.

5. R. R. Ernst and W. A. Anderson, *Rev. Sci. Instr.*, **37,** 93 (1966).

6. P. Felgett, Thesis, University of Cambridge, 1951; P. L. Richards *in* "Spectroscopic Techniques" (Ed., D. H. Martin). North Holland Publishers, Amsterdam 1967; R. Kaiser, *J. Magn. Res.*, **15,** 44, (1974).

7. N. Boden, *in* "Determination of Organic Structures by Physical Methods" (Eds, F. C. Nachod and J. J. Zuckerman). Academic Press, New York and London, 1971, Vol. 4.

8. R. Freeman and H. Hill, "FT NMR," Review article in "Molecular Spectroscopy", The Institute of Petroleum, London, 1971, 105.

9. H. Hill and R. Freeman, "Introduction to Fourier Transform NMR." Varian Associates A1D, 1970.

10. D. G. Gillies and D. Shaw, "The Application of Fourier Transformation to High Resolution NMR Spectroscopy," Ann. Reports on NMR Spectroscopy (Ed., E. F. Mooney). Academic Press, London and New York, 1972, Vol. 5.

11. W. Bremser, H. D. W. Hill and R. Freeman, *Messtechnik*, **79,** 14 (1971).

12. E. Breitmaier and W. Voelter, "^{13}C NMR Spectroscopy." Verlag Chemie, Weinheim, 1974.

13. R. N. Bracewell, "The Fourier Transform and its Applications." McGraw-Hill, New York, 1965.

14a. E. A. Guillemin, "Theory of Linear Physical Systems." Wiley Interscience, New York, 1963.

14b. R. R. Ernst, *J. Magn. Res.*, **1,** 7 (1969).

15. R. R. Ernst, "Sensitivity Enhancement in Magnetic Resonance," Advances in Magnetic Resonance. Academic Press, New York and London, 1957, Vol. 2:

16. I. J. Loewe and R. E. Norberg, *Phys. Rev.*, **107,** 46 (1957).

17. R. R. Ernst and R. E. Morgan, *Mol. Phys.*, **26,** 49 (1973).

18a. R. R. Ernst, *Chimia*, **26,** 53 (1972).

18b. R. Freeman and H. D. W. Hill, *J. Chem. Phys.*, **54,** 367 (1971).

19. J. S. Waugh, *J. Mol. Spectroscopy*, **35,** 298 (1970).

2

The role of the computer

We have seen that, basically, the advantage of the FT technique is the speed with which it can accumulate data over wide spectral regions. The words "speed" and "accumulate" are significant. These processes require the extensive use of a sophisticated storage facility; consequently, modern spectrometers involved in this sort of spectroscopy are generally equipped with a small dedicated computer. The computer functions in the combined roles as coordinator, data acceptor and mathematician. In many software packages it controls the timing of the measuring and decoupling pulse sequences, the acquisition of data and may even engineer the measuring sequences involved in the locking procedure. For our purposes, two of its most important functions are (1) the control and storage of data and (2) the translation of the spectral parameters sweep width and resolution into basic computer functions.

2.1 Data input

The first step in this chain involves the transfer of analog data from the receiver of the spectrometer to the computer memory. The problems associated with such an analog-to-digital conversion, that is, the change of an analog voltage to a storable digital number, have been considered in some detail.[1, 3] The magnitude of the number resulting from this digitization process directly governs how long one may accumulate data. It is sometimes thought that it is only necessary to continue running the experiment until sufficient signal-to-noise is obtained to yield a satisfactory spectrum.

In practice the actual number of scans which can be performed before the signal and/or the noise "overflows" the memory of the computer is determined by the word length of the computer. This question of overflow is also referred to as the dynamic range of the computer. The concept of word length, typically 16–20 bits in such dedicated computers, refers to the fact that the computer digitizer accepts the voltage coming from the receiver and converts it into some number $= 2^{n-1}$, where $n =$ the digitizer resolution (one bit is usually reserved for the sign of the signal). After n such conversions the magnitude of this number may be greater than the length of the word. In this case we say that the vertical memory of the computer has been exceeded (data overflow). This type of data destruction should be clearly differentiated from the case where any one incoming signal is too large for the *digitizer* to handle. Here it is only necessary to decrease the gain of one of the spectrometer amplifiers so that the analog voltage coming into the digitizer is of a manageable size. When the vertical memory has been exceeded the entire spectrum is destroyed, since each point in the free induction decay contains information from all frequencies. A simple, if somewhat unrealistic† example is instructive.

Assume:

1. The digitizer resolution is 12 bits.

2. The computer word length is 20 bits.

3. The incoming signal is one volt and represents a signal/noise (in the time domain) equal to one.

4. The receiver is set so that the incoming signal is full scale.

Question:

5. How many scans before the memory (biggest number which can be stored) overflows?

6. What will the increase in S/N be at this time?

Calculation:

$$2^{n-1} = 2^{12-1} = 2^{11}.$$

The signal adds coherently; the noise as $\sqrt{N_s}$; N_s = number of scans.

After 2^8 pulses total signal = 2^{19}.

Word length = $2^{20-1} = 2^{19}$ (signal will overflow if more are taken).

Noise increase = $\sqrt{2^8}$ = 2^4.

Total noise   = $2^{11} \cdot 2^4 = 2^{15}$.

Total signal   = $2^{11} \cdot 2^8 = 2^{19}$.

$$\begin{Bmatrix} \text{signal/noise} \\ \text{increase} \end{Bmatrix} = 2^{19}/2^{15} = 2^4 = 16 = \sqrt{N_s}$$

---

† For signal averaging with poor signal-to-noise, which is almost invariably the case, more experiments may be accumulated. See "The Computer in Fourier Transform NMR", J. W. Cooper in "Topics in ^{13}C NMR Spectroscopy" (Ed., G. C. Levy), Wiley Interscience, New York, 1976, Vol. 2.

Therefore, after 2^8 scans the memory overflows. The S/N will have increased by a factor of 16. If the signal enhancement is not sufficient the operator should consider, as one alternative, decreasing the digitizer resolution, thus permitting more experiments to be summed. Such a decrease (e.g. $12 \rightarrow 10$ or $12 \rightarrow 8$ bits) doesn't significantly alter the accuracy with which nmr signals are characterized.†

2.2 Sampling

It is useful for the reader to develop a feeling for how the computer characterizes an incoming frequency, since in the FT experiment an exponentially decaying series of sine waves is being monitored as a function of time. The computer samples the data for some time, t, in each of its memory locations. The time it spends in each of these is the dwell time.[2] The product of the dwell time and the total number of points utilized in collecting data is the acquisition time. The highest frequency which may be properly characterized by a computer sampling at frequency S is $S/2$, the so-called Nyquist frequency. In reality when the operator selects a spectral width for his experiment and enters this value via the teletype, the computer calculates immediately at what speed it must sample to properly characterize this frequency. The relationship between spectral width and dwell time has been shown to be:

$$\text{Spectral width (Hz)} = 1/(2 \text{ [dwell time]}).$$

Figure 2.1 demonstrates how the computer samples and characterizes. In Fig. 2.1a the frequency, X, is sampled at a rate greater than $2X$ ($\simeq 5$ pts/cycle) and thus is properly characterized by the computer. In Fig. 2.1b a frequency is shown which has been sampled twice during its cycle and represents, for this sampling rate, the Nyquist frequency. In the event that the sampling frequency is insufficient to correctly define a given sine wave, the points sampled will appear to the computer to have characterized a different, lower, frequency. This is demonstrated in Fig. 2.1c. The computer interprets the higher frequency $(S/2+X)$ as a sine wave with frequency $(S/2-X)$, and this is sketched, for clarity, in Fig. 2.1d.

Thus, a line with frequency $(S/2+X)$ *will appear* in the spectrum but at an incorrect position. Such a line is said to be folded. How can we recognize a folded resonance? One hint comes from the phase of this line relative to all the other lines in the spectrum. As will be shown shortly, there is usually a linear drift in the phase difference of each line across a wide spectral region.

† Recently, a number of FT nmr software packages have been updated such that the digitizer is reduced automatically and the FID divided by a suitable constant whenever data overflow is imminent. The measurement is then continued and the process repeated until the minimum digitizer resolution has been attained. At this time the program automatically halts thus preventing data destruction.

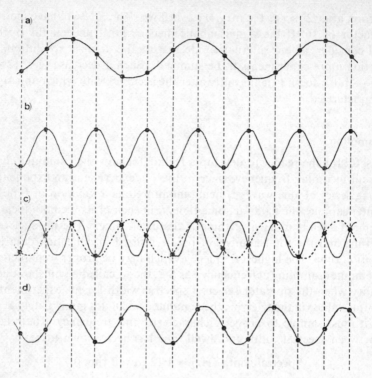

FIG. 2.1. Sampling the FID.

Since this line is not in its proper position its phase will be incorrect relative to the other lines in the spectrum. Additionally, most modern spectrometers have filtering devices incorporated near the input to the computer to cut down on the amount of noise which enters the input at frequencies higher than that of the Nyquist analog. Therefore, in addition to being incorrectly phased, the intensity of a folded line will generally be reduced relative to the non-folded resonances in the spectrum.

A second pertinent type of folding is associated with the pulsed frequency. Since the computer does not distinguish between frequencies which are higher or lower than this frequency ($+S/2$ or $-S/2$), commonly called the carrier wave, the operator customarily sets the carrier at one end of the spectrum to be measured.† There is no theoretical reason why the carrier cannot be placed in the middle of the spectrum since it is a simple calculation to reorder the spectrum in terms of high and low frequencies. Difficulties

† This problem has been eliminated using a new detecting system commonly called "Quadrature" detection. See Appendix D.

arise, however, for complicated previously unassigned spectra since the time involved in ascertaining which of the lines are actually carrier folded can be prohibitive.

Modern computers are commonly equipped with the ability to sample at rates of the order of 50 KHz. This is sufficient for the more "routine" nuclei, whose total frequency ranges are generally $<25\,000$ Hz. Even at frequencies corresponding to superconducting fields the sampling requirement for ^{13}C (67·9 MHz at 63 kG) will be approximately $(200\,ppm)$ $(67·9\,Hz/ppm) \approx$ 13·6 KHz which is well within the capability of the instrument; however, for ^{1}H and ^{19}F faster rates will be required.

2.3 Memory and measurement

How much computer memory is required for a routine experiment? To properly answer this question, we must first specify the nucleus to be studied since the observed line widths as well as the necessary spectral resolution vary from one nucleus to another. For ^{1}H Fourier, where the measurement of both small differences in chemical shifts and relatively small coupling constants are routine, the resolution requirements will be different than for ^{14}N Fourier, where the lines are broad and coupling to other nuclei seldom observed. Clearly, an increase in the number of memory points used to store n frequency units will result in increased resolution. Therefore a spectrum containing 1000 Hz placed in 2048 data points will produce an effective resolution of $(1000/2048) \times 2$. With twice as many data points the resolution will be correspondingly better. We should clearly distinguish between the resolving power of the spectrometer, which is a function of the homogeneity of the field, H_0, sample, shimming, etc. (which may be of the order 0·1 Hz, or less, for protons), and that which is created by selection of a given spectral width and size of memory. The first limits the attainable spectral resolution, whereas the latter represents a practical limit, which is set higher or lower as the experimental situation dictates. We shall see later (Chapter 3) that the size of the memory selected will also affect the time between the pulses and thus the equilibrium magnetization.

2.4 Data manipulation

In the course of the data accumulation, the single FID's will have been coherently added and stored in the digital computer. We now deal with the question of how these data are mathematically manipulated and eventually plotted as the well recognized frequency spectrum complete with numerical analysis. While the primary mathematical function is the actual Fourier transformation, there are several manipulations that can be performed on the

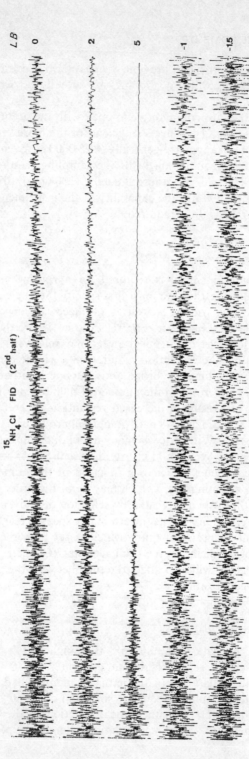

FIG. 2.2. The appearance of the FID as a function of TC, the time constant of the exponential ($TC = -LB\pi t_{acq}$). Only the second half of the FID is shown. (Courtesy of Spectrospin AG.)

FID before transformation which may improve the quality of the final spectrum.

A useful preliminary is the reproduction and separate storage of the FID. Not all subsequent arithmetic operations are reversible and such a storage operation may save many additional hours of measuring time should the transformed frequency spectrum be in some way unsatisfactory. Quite often it is only necessary to copy and transfer the FID from one data block to another. Where available, a disk or tape system can provide the memory space necessary for containing larger FID's.

From previous experience with continuous wave nmr spectroscopy we know that the S/N can be improved by filtering. Electronically this is performed by introducing a one or two step resistance-capacitance (RC) filter into the circuit.[3] With this arrangement frequencies are suppressed whose periods are shorter than RC. The optimum "band pass" for such a filter is that value which corresponds to the period of the nmr-signal. We can think of this as if the signal in the frequency domain were convoluted by a function which corresponds to the signal line shape. The completely analogous procedure in the time domain is the multiplication of the FID with an exponential function e^{-xt} (the Fourier transform of a Lorentzian line is an exponential function). In practice, each point of the FID is multiplied by $e^{ITC/N}$, where N is the memory size, I is an index varying from zero to $N-1$ and TC is the time constant equal to

$$TC = -LB\pi t_{acq}; \quad LB = \text{line broadening and } t_{acq} = \text{acquisition time.}$$

Thus the first data location of the FID is multiplied by e^0 ($I = 0$) and the last by e^{TC}. Whether we provide the program with a line broadening parameter, LB, or a sensitivity enhancement parameter equal to $\frac{1}{2}(t_{acq}/TC)$, the computer recognizes these data as the information required to calculate the TC value to be used. The smaller the time constant of this function, the larger is the sensitivity enhancement; however, the price we pay for this sensitivity, as in the CW experiment, is line broadening in the frequency spectrum. Conversely, if the weighting function is increasing with time, we can improve spectral resolution at the expense of sensitivity. The borderline case of these methods would be to weight the exponentially decaying FID with an exponential function increasing with the same time constant. "Sensitivity enhancement" as well as "resolution enhancement" are completely artificial processes and indiscriminate use of these may well result in detrimental broadening in the spectrum and/or loss of signals. This is especially true in cases where the spectrum contains lines of low intensity. Examples of the effects, on both the FID and frequency spectra, of varying the time constants from positive to negative values are shown in Figs 2.2 and 2.3.

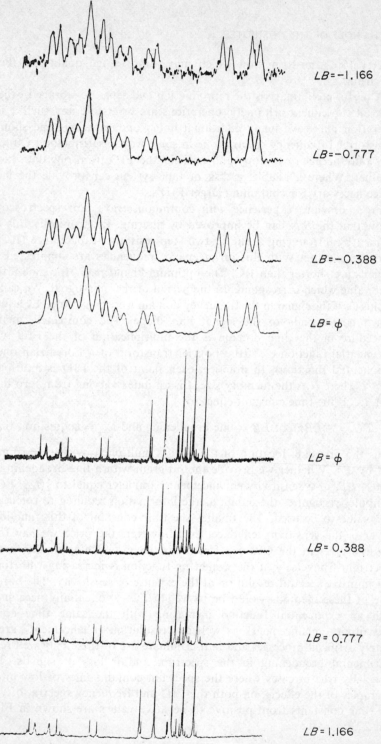

FIG. 2.3. The appearance of the frequency spectrum as a function of the time constant, TC ($TC = -LB\pi t_{acq}$). (Courtesy of Spectrospin AG, 1975.)

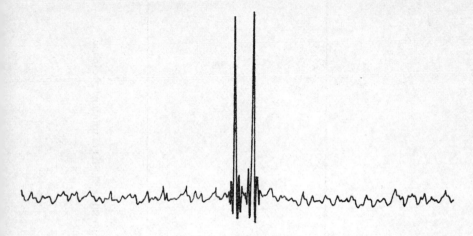

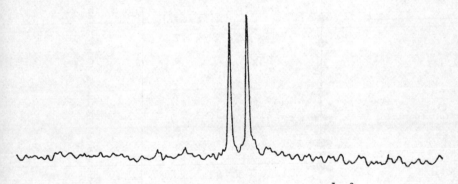

FIG. 2.4. The effect of truncation on the $^{13}C_1$ signal of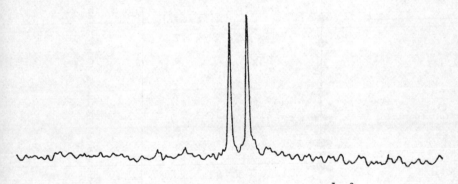

(a) Truncated FID and (b) FID mathematically driven to zero.

We have seen in the previous chapter that we collect the signal which is decaying in the receiver during the time, t_{acq}. It is possible, however, that the signal has not yet completely decayed during this period. Such a truncation of the FID, which is clearly illustrated in Fig. 2.4, is mathematically synonymous with multiplication of the spectrum by a function $(\sin \omega t)/\omega t$, the Fourier transform of a square wave. As a consequence of this condition, the signal exhibits lobes on both sides of the resonance (see Fig. 2.4) which might disturb the detection of weak lines in the neighbourhood of the truncated signal. This artificial digital broadening can, in part, be eliminated

C

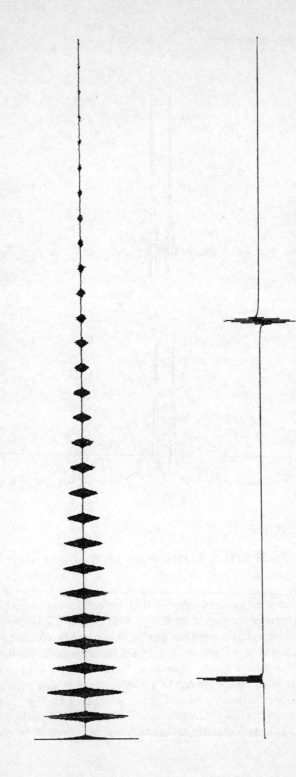

Fig 2.5. The two solutions (real and imaginary) resulting from the transformation of the ^{31}P FID of trimethyl phosphite.

by using an apodization routine[4] in which one multiplies the FID with a function $[\sin (\omega t)/(\omega t)]$.[2] Characteristically this function falls off more rapidly on both sides of the signal. Thus while the spectrum resulting from apodization will not be perfect it will be considerably improved.

Since the required Fourier transformation of the time domain function into the frequency spectrum is performed in a digital computer we do not calculate an integral of the form eqn (1.17), but rather sum over a series of finite data points. This means we do not perform a continuous, but rather a discrete FT. From the values $x_k(t)$ of the data at time t (this is the value of the FID) we determine $X_l(v)$ which is the coefficient of the Ith point in the frequency function

$$X_l(v) = \sum_{k=0}^{N-1} x_k(t) \cdot e^{-2\pi i v t/N}. \qquad (2.1)$$

The discrete FT is frequently performed using the Cooley–Tukey-algorithm.[5, 6] Instead of a simple multiplication and summation according to the equation above this algorithm functions via, the repetitive pairwise sorting of data points and therefore works best for a number of points, N, which is a power of 2 (2^N; in general 4096 or 8192). The calculation is performed "in place" on the data originally stored. In the course of the manipulation, the initial data is overwritten by intermediate values and finally by the resulting coefficients. Therefore the transformation of N data points requires only slightly more than N words of the core. The disadvantage lies in that the original information, the FID, is now lost (this is, in itself, sufficient motivation for performing the copying process suggested above). After the transformation the spectrum consists of $N/2$ real points and $N/2$ imaginary points corresponding to the cosine and sine transforms of the data. Since the two solutions are identical, except for phase (see Fig. 2.5), we finish with a spectrum which contains one-half of the points in the FID. The two solutions may then be displayed on an oscilloscope in order to permit phase correction.

The exponential factor $e^{-2\pi i v t}$ in eqn (2.1) can also be expressed as

$$e^{-2\pi i v t} = \cos 2\pi v t - i \sin 2\pi v t. \qquad (2.2)$$

Correspondingly, we can define the spectrum by a series of sine terms and a series of cosine terms. In the ideal case, the cosine series will then represent the pure absorption mode. In reality a mixture of absorption and dispersion modes occurs which leads to phase errors. A constant (i.e. not frequency dependent) error arises from misadjustment of the reference phase relative to the receiver phase detector. A phase error that varies linearly with time results from a delay between the end of the pulse and the beginning of data acquisition. Additionally, electronic filtering may influence the phase since a simple RC filter possesses a phase shift. In order to eliminate the mixing of modes in the final spectrum the coefficients $X_l(v)$ of the spectrum are calculated from two new sine and cosine series (S_1 and C_1), which represent weighted average values of the original series

$$X_l(v) = K.C_1 + (1-K^2)^{\frac{1}{2}}.S_1.$$

In this method the factor K is regarded as a linear function of the frequencies.† In principle the correct K can be calculated on the basis of an iterative procedure through the computer. Generally, in the commercially available spectrometers, this is performed in such a way that the operator himself optimizes K by inspecting the spectrum on the oscilloscope. This is commonly done with two phase knobs which directly alter the value K in the above relation.

The spectrum after optimization using the phase correct routine, can now be transferred from the computer to the recorder of the spectrometer. The plot-parameters, such as vertical scale, the starting and final frequency and the value of the desired offset can be introduced via the teletype. Starting and final frequencies, which are given for an expansion of the original spectrum, must, of course, be within the total spectral width. For all signals whose voltages are higher than a given threshold intensity, the computer will, on command, calculate and print frequency positions, intensities and, given the correct spectrometer frequency and a suitable reference, chemical shifts in ppm thus minimizing subsequent operator data handling.

References

1. R. R. Ernst, *J. Magn. Res.*, **4**, 280 (1971).
2. S. Goldman, "Information Theory." Prentice Hall, Englewood Cliffs, New York, 1953.
3. R. R. Ernst, "Sensitivity Enhancement in Magnetic Resonance," Advances in Magnetic Resonance. Academic Press, New York, 1966, Vol. 2.
4. R. Bracewell, "The Fourier Transform and its Applications." McGraw-Hill, New York, 1965.
5. J. W. Cooley and J. W. Tukey, *Math. Comput.*, **19**, 296 (1965);   R. Klahn and R. R. Shively, *Electronics*, **124**, (1968).
6. W. T. Cochran, *et al.*, *Proc. IEEE*, **55**, 1964 (1967).

† A phase error that cannot be corrected using such a routine stems from imperfections in the pulse. In this situation we recommend a discussion with your manufacturer.

3

Spectrometer operation

In this section we deal with the techniques involved in measuring a "normal" FT spectrum for a routine nucleus. In addition to the necessity of distinguishing between the pulsed and CW techniques, we should remember that usually the experiment is a multiple resonance one. This will call for separate transmitter and receiver circuitry for each of the nuclei involved. Thus, for a *proton*-decoupled ^{31}P (^{15}N, ^{13}C ...) spectrum with *deuterium* lock there will be *three* frequencies to be regulated. We shall now consider these individually.

3.1 Locking

We have seen that one of the important functions of the computer involves the repetitive storage of data. Obviously the incoming signals must be repeated as exactly as possible so that the end result will be a sharp, readily interpretable spectrum. A drift in H_0 will result in a change in the frequencies of resonance, v, such that data accumulated at time t will have resonance frequency v, whereas at time $(t + \Delta t)$ these same nuclei will have resonance frequency $(v + \Delta v)$. The spectrum which results after long term averaging will be less than satisfactory in that the lines will be broadened.† To prevent such a drift in H_0 the operator will routinely "lock" the spectrometer. The term "lock" is used in that the researcher actually performs a separate nmr experiment on a nucleus within the sample tube and then uses an electronic circuit to monitor whether there has been a drift in the frequency of resonance of this signal due to a change in H_0. This is readily envisioned if one remembers that it is possible to measure the nmr signal in the so-called dispersion mode (e.g. Fig. 3.1a). In this mode at the exact resonance frequency, v, the signal has zero intensity. A drift in the field, H_0, according to eqn (3.1), reproduced here,

$$v = H_0 \gamma (1 - \sigma)/2\pi, \tag{3.1}$$

will simultaneously produce a change in the resonance frequency of our lock

---

† Where the lines are already quite broad, e.g. quadrupolar nuclei, locking may not be necessary since the drift in H_0 could be $\ll$ line width.

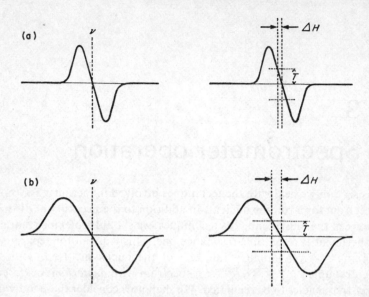

FIG. 3.1. Locking on signals of widely differing line width. (a) A relatively sharp line and (b) a relatively broad line.

substance. If we have previously balanced our circuit at this null, the drift in H_0 will result in a signal (be it positive or negative) where previously there was none. The electronics can then "rebalance" our system by applying such current to the magnet as is necessary to return the field to its previous value. Thus we have effectively fixed or "locked" the field to that position corresponding to the resonance frequency of our lock substance. Naturally we select as a lock substance a material which affords a sharp line. Most detecting systems have a finite threshold voltage which must be exceeded before the system begins to "correct". It may be seen clearly from Fig. 3.1b that, with a lock substance having a broad resonance, the field will drift further before reaching signal intensity T (our so-called threshold voltage) than it will for a lock substance with relatively sharp line (almost no drift at all).†

Current commercial FT spectrometers come equipped routinely with a multi-nuclear lock capability. The most frequently used lock nucleus is undoubtedly deuterium, ^2H, although ^{19}F and occasionally, when broad band proton decoupling is not required, ^1H are still employed. The choice of deuterium is related directly to the routine use of deuterated solvents for ^1H nmr studies and their consequent ready commercial availability.

† In reality modern spectrometers employ a locking system which continuously monitors the stabilization signal.

The selection of the lock substrate is not necessarily a trivial decision for ^1H nmr. We know that saturation of the lock signal will broaden its resonance. Since the ease with which deuterium atoms in various chemical situations will saturate is not constant, the operator should consider the selection of a lock material in terms of the aims of the experiment (e.g. high resolution ^1H studies in which coupling constants of magnitude $\ll 1$ Hz are to be measured are probably best *not* studied in $CDCl_3$ which has rather a broad deuterium resonance). Acetone-d_6 and benzene-d_6 are amongst those solvents whose ^2H resonances are reasonably sharp. Part of the problem involving chlorinated lock materials stems from the possibility of isotope effects (^{35}Cl and ^{37}Cl) on the position of the lock resonance. This is valid for both $CDCl_3$ and $CFCl_3$ (a potential ^{19}F lock material). In addition, most deuterated solvents are both chemically and isotopically impure; thus, commercial samples of $CDCl_3$ will frequently show a proton signal due to water† somewhere near $1\cdot5$ ppm. For single passage spectra this signal is not usually observed; however, it is quite visible in ^1H spectra which have been accumulated and might well obscure an important resonance. Additionally, the $CHCl_3$ in the deuterated solvent may well be the largest signal in the ^1H spectrum of dilute solutions with the result that meaningful integration in the aromatic region could prove difficult.

The problem of isotopic purity is especially severe in the study of aqueous biological systems. The long troublesome HDO signal, when accumulated, can lead to difficulties both with line overlap and dynamic range. A number of technical suggestions aimed at the solution of this latter problem have been made and we will consider these in Chapter 5.

In connection with the above considerations, it should be remembered that, since the ^2H nucleus is quadrupolar, there may be a marked temperature dependence on the width of the lock resonance. Even relatively sharp ^2H resonances may broaden considerably below temperatures of the order of 180–190°K thus further complicating low temperature measurements. In such situations a fluorine lock from a suitable Freon molecule may prove useful.

3.2 Decoupling

While most nmr spectrometers, at all levels, can be equipped to perform double and multiple resonance experiments, this work is sometimes thought of as "special" requiring accessories to the main instrumental components. In the observation of many X nuclei ($X \neq {}^1H$) decoupling techniques have become routine and it is therefore useful to review some basic concepts of double resonance before passing on to practice.

† This may be removed by normal drying procedures.

General description

In a double resonance experiment $A\{X\}$† the nuclear spins are under the influence of not only the static and "measuring" fields, H_0 and H_1, respectively, but a third decoupling field, H_2, as well. This latter disturbs selected nuclei within the molecule at the same time that H_1 induces the transitions which we measure as our spectrum.

We may easily envisage the result of this secondary disturbance if we return conceptually to the rotating frame of reference with, in this case, the coordinate system rotating around the z axis with a frequency equal to that of the decoupling frequency ω_2. In analogy with Chapter 1, eqn (1.13) when the RF field H_2 is at the resonance frequency of X with sufficient strength the fictitious field ω_2/γ cancels the stationary field H_0 and the effective field for the decoupled nucleus is H_2. Consequently, the direction of quantization for these spins is along the x axis. However, the direction of quantization for the A nuclei *remains* the z axis (these nuclei are still preceding, around the z axis). Since the coupling between two nuclei, A and X, is a scalar product

$$A \cdot X = A \cdot X \cdot \cos \theta \tag{3.2}$$

where θ is the angle between the quantization axes, the two nuclei will be "decoupled" since $\cos \theta = 0$ (x and z are 90° apart).[1]

The general decoupling phenomena[2, 3] are usually subdivided according to the strength of H_2 as follows.

1. $\gamma H_2 \gg J(A, X)$. This will result in the collapse of multiplet(s) connected to resonances at, or close to, the irradiating frequency and represents a common form of spin decoupling.

2. $\gamma H_2 \approx J(A, X)$. The collapse of a specific multiplet or selected decoupling.

3. $\gamma H_2 = \Delta v_{\frac{1}{2}}$. ($\Delta v_{\frac{1}{2}}$ equals the line width at half height.) The so-called spin tickling experiment.

4. $\gamma H_2 \ll J(A, X)$. The Nuclear Overhauser Effect (NOE). In this case intensity changes in the A resonance are observed by saturating transitions in the X spectrum.

Decoupling techniques

Complete decoupling may present experimental difficulties when the nuclei to be irradiated cover a large frequency range. The problem has been over-

† We adopt here the common notation of $A\{X\}$ where X is the nucleus perturbed and A the nucleus observed.

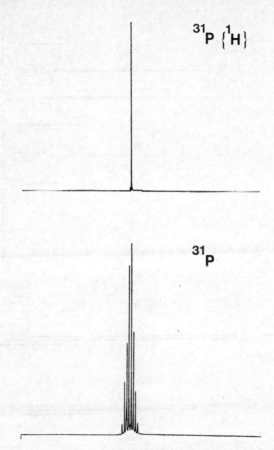

FɪG. 3.2.   The frequency ^{31}P spectrum of trimethylphosphite with and without ^{1}H decoupling.

come via a form of broadband decoupling[4, 5] in which the so-called "noise-modulation" technique is used. In this method the field H_2 is applied pseudo-randomly using a pulse-shift generator which quickly generates all the relevant frequencies from the single CW frequency. The output of this generator is then amplified and applied to the probehead. With the decoupling carrier frequency appropriately set, all of the X nuclei are simultaneously decoupled producing sharp signals in the A spectrum (see Fig. 3.2)†.

The complete decoupling of X from A usually produces an A spectrum with better sensitivity; however, this is done at the expense of the consider-

† For $X = {}^1$H the selected setting usually lies in the middle of the ^{1}H spectrum. For other nuclei (e.g. ^{19}F) the frequency range of the chemical shift may be too large to affect complete decoupling with any one particular setting.

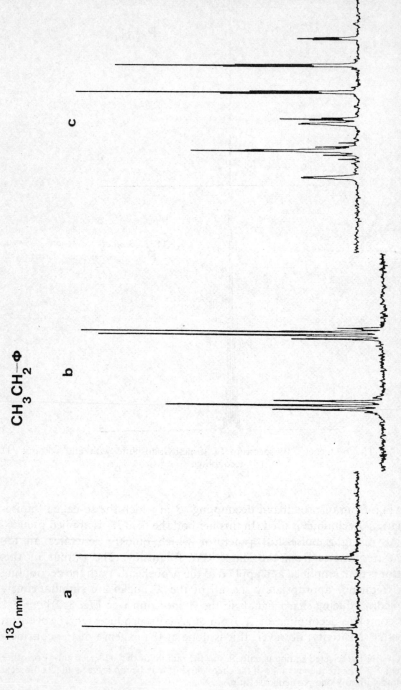

$CH_3CH_2-\Phi$

^{13}C nmr

a

b

c

FIG. 3.3. The aliphatic section of the ^{13}C spectrum of ethylbenzene under the condition of (a) complete ^1H decoupling, (b) off-resonance ^1H decoupling and (c) complete " coupling ". [$(CO_3)_2$ CO observable.]

able information (structural, configurational . . .) which may be extracted
from a knowledge of the magnitude of $^nJ(A, X)$. A useful compromise, for
the case where $X = {}^1H$, involves irradiation with a relatively strong CW H_2
field somewhere "off-resonance". The resulting spectrum will frequently
retain a reduced value of $^1J(A, {}^1H)$ but no long range (e.g. $^nJ(A, H)$, $n > 1$)
couplings. This situation and the two extremes, complete (broad band)
decoupling and "coupling" is shown in Fig. 3.3 for the carbon resonances
of the ethyl group in ethylbenzene. The off-resonance decoupling technique
is of especial value in the assignment of carbon-13 spectra since it readily
distinguishes between methine, methylene and quaternary carbon resonances.
Additionally, for cases where the proton spectrum is known, it is possible to
utilize the residual value of $^1J({}^{13}C, H)$ for assignment purposes. It has been
shown that

$$^1J({}^{13}C, H)_{residual} = {}^1J({}^{13}C, H)\Delta v/2\pi\gamma H_2 \tag{3.3}$$

where Δv is the difference between the resonance frequency and the irradiating
frequency, $^1J({}^{13}C, H)$ is the full value of the one bond coupling, and H_2 is
the strength of the decoupling field.†

The specific decoupling experiment, that is, $\gamma H_2 \approx J(A, X)$, has retained
its classical importance. In addition to the $^{13}C\{H\}$ experiment which permits
line assignment in the ^{13}C spectrum via irradiation in the proton spectrum
at a specific position, $^{13}C\{^{31}P_{sel}{}^1H_{bb}\}$ (broad band proton decoupling,
selective phosphorus decoupling) has been found to be useful in simplifying
complex carbon spectra (see Fig. 3.4[6]). Of course, selective decoupling implies
an exact knowledge of ^1H (or ^{31}P) nmr spectrum. An interesting variant
for tying together the ^1H and ^{13}C spectra, when the ^1H spectrum is not
known has been described.[6a]

Signal intensity and NOE

In addition to the collapse of multiplets, broad band decoupling may lead
to a further signal enhancement resulting from changes in the spin populations
(and thus in signal intensity). Such a phenomenon, when induced by double
resonance, is generally referred to as nuclear Overhauser enhancement
(NOE).[2, 7]

Let us consider the energy level diagram for two spin $= \frac{1}{2}$ nuclei, A and X,
which are not J-coupled. As shown in Fig. 3.5, this consists of four levels,
whose populations we shall call p_1–p_4.

In general the possible relaxation probabilities are:

W_1^A The probability of the transitions $2 \leftrightarrow 4$ or $1 \leftrightarrow 3$ (single quantum
transition).

W_1^X The probability of the transitions $3 \leftrightarrow 4$ or $1 \leftrightarrow 2$ (single quantum
transition).

† For this equation to be valid $\gamma H_2/2\pi$ must be $\gg \Delta v$, J.

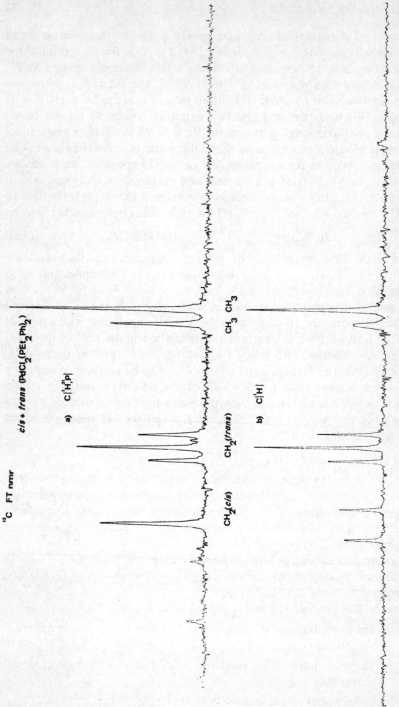

FIG. 3.4. Aliphatic part of the ^{13}C spectrum of a mixture of *cis*- and *trans*- [PdCl$_2$(PEt$_2$ Ph)$_2$]. The triplet for CH$_2$(*trans*) arises from "virtual coupling" phenomena. Spectrum (a) was measured with simultaneous irradiation of the ^{31}P resonance of the *cis* isomer thus allowing assignment of the spectrum. Spectrum (b) was measured without ^{31}P decoupling.

W_2 The probability of the transition $1 \leftrightarrow 4$ (double quantum transition which has significance when dipole–dipole relaxation is
important).

W_0 The probability of the transition $3 \leftrightarrow 2$ (zero quantum transition).

If $X = {}^1H$, the broad band decoupling experiment will saturate the transitions
$2 \leftrightarrow 1$ and $4 \leftrightarrow 3$ and equalize the population of these states (e.g. $p_1 = p_2$,
$p_3 = p_4$). For simplicity we assume that $p_2 \approx p_3 = C$, so that we may write
simple expressions for the populations before and after broad band irradiation. We take Δ to be the difference in population

$$\text{Before: } p_1 = C - \Delta \qquad \text{After: } p_1 = C - \Delta/2$$
$$p_2 = C \qquad\qquad\qquad p_2 = C - \Delta/2$$
$$p_3 = C \qquad\qquad\qquad p_3 = C + \Delta/2$$
$$p_4 = C + \Delta \qquad\qquad p_4 - C + \Delta/2$$
$$p_4 - p_1 = 2\Delta \qquad\qquad p_4 - p_1 = \Delta$$

between p_1 and p_2 as well as p_3 and p_4 with the lowest energy state more
populated. Clearly, the populations of p_1 and p_3 are increased at the expense
of p_2 and p_4 although the sum of all the populations remains constant. Since
the gain in the populations p_1 and p_3 and the decrease in p_2 and p_4 are the
same, the W_1 processes do not produce an NOE. However, the W_2 process
(cross-relaxation) acts to increase p_4 and decrease p_1 while restoring the
population difference $p_4 - p_1$ to its equilibrium value. This is just what is
required to enhance both A transitions. Unfortunately, the W_0 transition
$3 \leftrightarrow 2$ tends to oppose W_2.

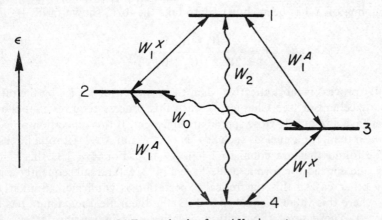

Fig. 3.5. Energy levels of an AX spin system.

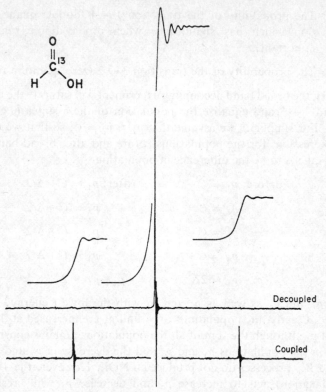

FIG. 3.6. ^{13}C spectra of formic acid increase in signal intensity due to NOE.[9a]

The fractional enhancement, f, of the signal of A with and without strong decoupling of X has been shown[7, 8] to take the form shown in (3.4).

$$f = \frac{W_2 - W_0}{2W_1{}^A + W_2 + W_0} \, (\gamma_X/\gamma_A).$$ (3.4)

The W_2 process is most effective when these nuclei relax (T_1) by the dipole–dipole mechanism (see Chapter 4) since this process requires that the two nuclei have a relatively close spatial arrangement. If this relaxation process is *dominant* then, for a nucleus such as ^{15}N or ^{13}C and $X = {}^1$H, total intensities can be found that are equal to $1 + [(\gamma_H/\gamma_A)/2]$. For $A = {}^{13}$C this value is 2·988 and indeed for formic acid,[9a] see Fig. 3.6 (and subsequently a great many other cases), this theoretical expectation is confirmed. Similarly for ^{15}N, where this value is equal to $-3·93$, the theoretical maximum has been observed. The case of ^{15}N, with a negative γ, is interesting in that, since several relaxation mechanisms may contribute, the possibility for nuclear Overhauser *nulling* of the signal exists and has in fact been found.[6, 9b]

Gated decoupling

While certainly a welcome addition when sensitivity is a problem, nuclear Overhauser enhancement introduces the problem of intensity ambiguity. Not every nucleus is dominated by the dipole–dipole relaxation mechanism and, therefore, the NOE may vary from carbon to carbon (or from nitrogen to nitrogen). Additionally, for those occasions when "coupled" spectra are to be measured, it would be useful if we could retain some or all of the NOE in the favourable cases.

These situations have been handled using so-called alternately pulsed or gated decoupling techniques.[10] By "gated" we mean a computer controlled sequence which switches the decoupler "on" and "off" at various intervals. We know that, while the acquisition time is often of the order or 1 s or less, T_1 values (^{13}C, ^{15}N ...) are of the order of seconds. Thus the time during which the distribution of spins due to the nuclear Overhauser effect approaches equilibrium can be significantly longer than t_{acq}. Therefore, if we apply decoupling power only throughout the length of the acquisition time, the result will be a decoupled spectrum free of nuclear Overhauser enhancement. We assume that in such a short period (< 1 s) little or no NOE develops. After the decoupling, a time, t_{post}, may be introduced to ensure that any very small

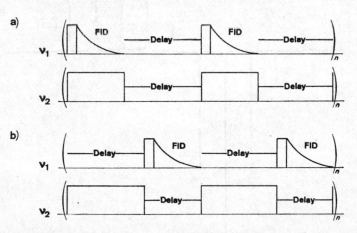

FIG. 3.7. The timing of the pulse sequences in (a) a decoupled spectrum without NOE and (b) a "coupled" spectrum in which the NOE is retained (lengths of ν pulses are not drawn to scale; ν_1 = measuring frequency, ν_2 = decoupling frequency).

NOE which might be present does not sum over a long period. Conversely, we may turn "on" the decoupler for some time t_{post}, during which the NOE builds up, quickly switch it off and then measure (pulse H_1) the spectrum. Immediately after removal of the decoupling the multiplicity due to the protons is restored, but the NOE, which is associated with the T_1 process, persists. Thus, the spectrum will contain nuclear Overhauser enhanced "coupled" signals. These two sequences of pulses are diagramatically sketched in Fig. 3.7.

A more selective form of gated decoupling has recently attracted some interest.[11, 12] In this variant the decoupler frequency, v_2, is positioned exactly at one absorption in the X part of an AX_n spectrum with an H_2 field strength,

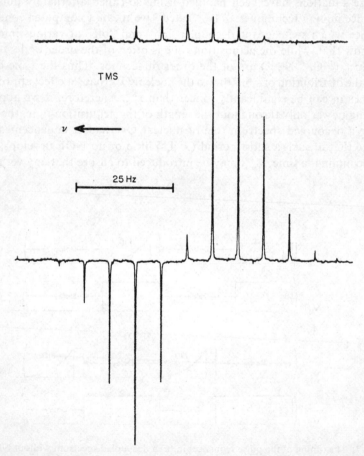

FIG. 3.8. Selective population transfer experiment for the ^{29}Si resonance of TMS demonstrating sensitivity enhancement.[13] Upper trace represents the normal spectrum.

$\gamma H_2/2\pi$, of the order of the line width. The decoupler is now switched "on" during a time, τ_p, *before* the measuring pulse such that $\gamma H_2 \tau_p = 180°$. Assuming relaxation to be negligible during the time of the pulse ($\tau_p \ll T_1$) complete inversion of the energy level populations may occur.† Such a pulsed decoupling technique, called selective population transfer, when followed by the normal analytical pulse, can afford a spectrum (Fig. 3.8) in which the *sensitivity* is considerably improved.[13] In addition to improving the signal-to-noise it has been shown that this technique can provide the same type of information with regard to relative signs of coupling constants as other forms of double resonance.

The combination of this latter method with "difference spectroscopy" (see Chapter 5) permits the display of an FT-spectrum which shows only the net effect due to selective population transfer. This spectrum results directly from the subtraction of an FID obtained without application of a π pulse, from that measured just after its application. Such a subtraction is readily performed in the computer. It should be pointed out that selective population inversion is also possible in the homonuclear mode,[14, 15] thus providing the analogous information in FT nmr spectroscopy as does the INDOR technique in the CW mode. In our previous classification of decoupling experiments according to the H_2-power we have not considered the question of whether the RF-fields H_1 and H_2 are stationary. A non-stationary (frequency swept) decoupling field is characteristic of the so-called INDOR-technique. In this method the intensity of a specific line is recorded while the decoupler frequency is swept over the range of the spectrum. A change in intensity of the monitored line occurs when ν_2 hits a resonance that has an energy level in common with the observed transition.

3.3 Measuring the spectrum—practical considerations

In the following section, we address ourselves to the procedures involved in measuring a spectrum using a "conventional" FT nmr spectrometer. For the most part we constrain ourselves to what we consider as "routine" operation, i.e. the procedures necessary to obtain a spectrum, in FT mode, with sufficient S/N to allow the scientist to extract the line positions and separations which he is seeking. Comments concerning more advanced experiments will be presented in a subsequent section. The approach shown here is a general one and is not directed towards any one brand of instrument, since the principles are common to all such instruments. A block diagram showing the basic components of an FT nmr spectrometer, as well as the paths taken by the various frequencies, is shown in Fig. 3.9.

We have assumed that the reader has some experience in measuring nmr

† Complete inversion has been shown for the AX system $^{13}CHCl_3$ but is not always the case.

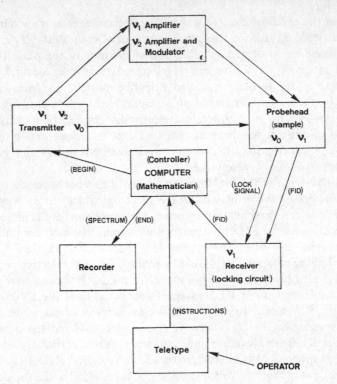

FIG. 3.9. Block diagram showing the major components of a Fourier nmr spectrometer as well as the flow of the individual frequencies.

spectra (i.e. Perkin–Elmer, R12, Varian T60 or equivalent) so that the widely used terms, "spectral width", "offset", "high field", etc. invoke the correct mental image. We will assume the sample has been properly prepared in advance. Thus it will most probably contain the following.

1. A lock source, perhaps from a deuterated solvent.

2. A reference material (for ^{13}C this may be the solvent itself).

3. A vortex stopper (necessary when large sample tubes are employed in order to prevent vortexing from disturbing the homogeneity across the sample).

4. The correct volume of solution; for 10 mm tubes this is usually of the order 2–3 cm^3, although specially constructed cells which constrict the volume of the sample to the receiver coil are commercially available.

The possibility now exists to routinely measure nmr spectra in sample tubes whose diameters vary from 5–15 mm,† and the operator should consider this in conjunction with the question of sample concentration. The usual requirement that the solution be devoid of small particles (undissolved solute, drying agent used to remove water from deuterated solvents, etc.) is still valid.

Locking

Insert the sample using the height gauge to properly position the solution in the receiver coil.

Locate the lock signal on the oscilloscope. This may require increasing the H_1 field *for the lock channel* (remember each channel; analytical, lock and decoupling will have associated with it an H_1 field).

Reduce the width of the sweep while maintaining the lock signal on the oscilloscope. This will generally entail making a small change in H_0.

If necessary shim crudely—the characteristics of the previous sample may have been considerably different.

Reduce the strength of the H_1 field to avoid saturation. Saturation will broaden the lock resonance.

"Lock" the spectrometer‡ (Depress the nmr stabilization button).

Shim on the spinning homogeneity controls. This is usually accomplished by maximizing the amplitude of the lock signal as it is read from either a meter or an oscilloscope.

Select the position of the carrier wave

At this point it is useful to recall that the computer does not distinguish between frequencies which are higher or lower than the carrier. Normally the carrier is set at one end of the spectrum so that a routine spectrum will have no carrier wave folded lines (just carrier wave folded noise!). Quadrature detection§ has been used to avoid this problem. However, it is frequently desirable to measure using a reduced spectral width and for these spectra folding (both types, see Chapter 2.2) can be annoying. In most cases it is useful to first measure the spectrum using a wide spectral width with the carrier at one end. Once all the resonance frequencies are known the position of the carrier and the spectral width may be adjusted such that, although folding occurs, the spectrum may be readily interpreted. Remember that

† A new commercial system has the capability of measuring spectra in 25 mm tubes.

‡ With most spectrometers steps 2–5 can sometimes be omitted. This is especially true when the same lock substance is used repetitively. In these situations after sample insertion the spectrometer will lock immediately the nmr stabilization button is depressed; however this is not always the case and it is useful for the novice to complete the entire sequence.

§ Quadrature detection refers to a relatively new detecting system now incorporated into several commercial spectrometers. The carrier wave is placed in the centre of the spectrum. See Appendix D.

frequencies higher than the Nyquist frequency, $S/2$, by some amount X Hz may appear as lines of reduced intensity with incorrect phase at $((S/2)-X)$ Hz. In connection with reduced intensities due to filtering, it is a good idea to choose the carrier wave position such that not all the resonances fall near the Nyquist frequency. Most spectrometer filters are not linear and lines near this frequency may be reduced in intensity.

We should recall that a change in lock substance will shift the entire spectrum and effectively may be thought of as a change in carrier wave offset. The actual frequency shift is represented by eqn (3.5),

$$\Delta(\text{offset}) = (\gamma_X/\gamma_{\text{lock}})\Delta\delta(\text{Hz}), \tag{3.5}$$

in which the frequency change in the offset is equal to the product of the frequency change in going from lock substance $A \rightarrow B$, $\Delta\delta$, and the ratio of the gyromagnetic constants for the measured nucleus, γ_X, and the lock substance, γ_{lock}.

A sample calculation is instructive:

Question:

What offset change in the ^{31}P spectrum will be necessary to compensate for a change in ^{2}H lock substance when acetone-d$_6$ is substituted for benzene-d$_6$?

The chemical shift difference for ^{2}H will be approximately the same as for ^{1}H, i.e.

$$\delta CD_3 \, COCD_3 \sim 2 \cdot 1$$

$$\delta C_6 \, D_6 \qquad \sim 7 \cdot 1$$

$$\Delta\delta \sim 5 \, \text{ppm} = (\text{for } ^{2}\text{H at 21 KG}) \, (5 \, \text{ppm})(13 \cdot 8 \, \text{Hz/ppm}) = 69 \, \text{Hz}$$

$$\gamma/2\pi \, (\text{Hz/Gauss}) \, ^{2}\text{H} = 653 \cdot 6$$

$$\gamma/2\pi \, (\text{Hz/Gauss}) \, ^{31}\text{P} = 1723 \cdot 5$$

$$\Delta(\text{offset}) = \left(\frac{1723 \cdot 5}{653 \cdot 6}\right) (69) \approx 182 \, \text{Hz}.$$

Since for shifts of the lock resonance to higher field the carrier frequency should be increased, we will have to add ~ 182 Hz to the offset frequency to maintain line positions on our chart paper.†

Selection of the decoupling mode

The operator now chooses the type of decoupling (if any) which is correct

† This assumes no solvent effect on the line positions, which is most probably not true.

for his experiment. This will frequently be broad band proton decoupling for routine ^{13}C and ^{31}P spectra, although the "off-resonance" technique has become routine for both these nuclei.†

In all cases, the position, H_2 strength and mode of decoupling (e.g. modulated CW, pulsed) will have to be chosen in accord with the principles of the previous section. For a routine ^{13}C spectrum with broadband ^{1}H decoupling one might select the following parameters.

1. Decoupling offset = set in the middle of the proton spectrum.

2. Broadband amplifier = 5–10 watts output power (this should represent the *difference* between the forward power and the reflected power as read by a suitable watt meter).

3. Decoupling mode = broadband (modulation).

It is worth remembering that the mechanics of *homo*-decoupling will be somewhat different for the FT experiment. The decoupling frequency may not be "on" (CW) during the acquisition of the data since the receiver coil will constantly "see" this relatively high power signal and transmit it to the computer. Thus it will be necessary to gate the decoupling frequency and such other components (i.e. mixing of frequencies) as are used in the experiment. This can be done by software control with, in some cases, the addition of gating devices. In the hetero-decoupling experiment (i.e. ^{13}C{^{1}H}) the decoupling is *not* "seen" by the receiver, because the latter is tuned to accept a different frequency and so the decoupler may be left "on" continuously. The operator must be aware of these differences especially since recent studies[16] have shown that pulsed decoupling and CW decoupling are not always completely analogous.

Spectral width and memory size

For nuclei with "relatively poor sensitivity",‡ the total time necessary to achieve a satisfactory spectrum is dependent on the sample concentration and nmr characteristics of the nucleus, in particular T_1 and T_2, the spin–lattice and spin–spin relaxation times, respectively. Since the selection of the title parameters may be based not only on the chemical requirements, but on T_1 as well, each factor must be considered for every experiment. The available

† The detection of hydride species, *in situ* via ^{31}P nmr studies of metal complexes containing phosphines may be achieved using off-resonance ^{1}H decoupling. Since the hydride chemical shift is at extremely high field (often $\tau = 20$–30 ppm) the ^{31}P spectrum will frequently show only a residual value of $^2J(P, H)$ with all other ^{1}H resonances decoupled.

‡ This is conveniently defined as the ratio of the product: (natural abundance) × (magnetic moment) for nucleus X, relative to that for ^{1}H. This should not be confused with spectrometer sensitivity.

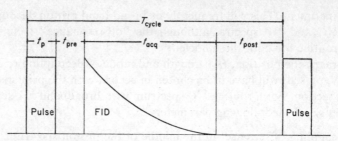

FIG. 3.10. Nomenclature for the various aspects of a single cycle in a multipulse experiment.

variables and their time relationship in the multipulse experiment are shown in Fig. 3.10 (not drawn to scale).

The selection of the spectral width (actually the dwell time) and the size of the memory is generally dependent on the type of information sought. Thus a proton spectrum will usually require ~1000 Hz with a resolution of at least 0·5 Hz. For ^{13}C the analogous values 5000 Hz and 1–2 Hz are good starting points. We have observed, that for ^{13}C, ^{31}P and ^{15}N a spectral width of 5000 Hz in conjunction with an FID containing 8 K (8192) points serves as a comfortable place at which to begin. This allows the operator an effective resolution of $\approx 2·4$ Hz. Once a preliminary spectrum has been obtained, the researcher may alter these spectral parameters to suit his chemistry.

If one accepts these values then we have defined an acquisition time as follows:

$$\text{spectral width} = 1/(2 \times \text{Dwell time}),$$

$$t_{acq} = N(\text{Dwell time}),$$

$$t_{acq} = N/(2 \times (\text{spectral width})),$$

$$N = \text{number of points in the FID},$$

from which it is clear that the time for a single experiment, $\approx t_{acq}$, will depend directly upon the size of the memory chosen to contain the FID and *inversely* on the width of the spectrum. Since increasing the number of data points in the spectrum, while holding the spectral width constant, brings an increase in spectral resolution we have the situation, in analogy with CW nmr, that speed and resolution are related inversely. The reader who is unfamiliar with these common FT nmr parameters will find Table 3.1 useful.

Selection of the pulse length and T_{cycle}

We recall from the introductory chapter that, in the rotating frame of

TABLE 3.1

Some computer parameters and their relation to the final spectrum

Dwell Time	Spectral Width (Hz)
50 μs	10 000
100	5 000 (commonly employed for ^{13}C, ^{15}N, ^{31}P)
200	2 500
250	2 000
500	1000 (commonly employed for ^1H)

Dwell Time	S.W.(Hz)	Memory Size FID	Hz/ channel[a]	No. transients/h
100	5000	8K	1·22	4394
100	5000	16K	0·61	2197
500	1000	8K	0·24	879
500	1000	16K	0·12	439

[a] At a field stress of $\approx$ 20 KG
[b] There are only $N/2$ points in a spectrum which utilizes N points for the FID

reference, the magnetization vector is tipped away from the z direction by an angle, $\alpha = \gamma H_1 t_p$. In current spectrometers, H_1 is usually held constant and so the proportionality between α and t_p is direct. After the pulse there is a time period (equal to the acquisition time when $t_{post} = 0$) during which the z magnetization decays exponentially towards its original position in the z direction with a characteristic time constant T_1. In the multipulse experiment a second pulse of equal length is applied before it reaches this value. If the value of t_p corresponds to 90°, the projection of the magnetization vector in the y direction will be smaller after the second pulse than after the first pulse. After n pulses of this type the signal in the receiver coil will *not* be building and the net signal induced may even approach zero since the negative contributions will cancel the positive ones. Naturally, if the time between the pulses was sufficiently long then the maximum signal could be had after each pulse. In practice this requires a long waiting period since for the signal to have returned to its original position will require a time longer than three T_1's. Thus the value selected for t_p should be somewhat less than 90° and depend on the longest T_1 value in the sample (if all the signals are to be observed) and the time between the pulses (see Chapter 1). A simple example is shown in Fig. 3.11. In practice the choice of pulse length, t_p, is generally decided empirically, since, in most cases, the spin lattice relaxation time, for the compounds under consideration is not known. However, if one can qualitatively estimate T_1 then it is possible[17a] to select a suitable value for the flip angle,

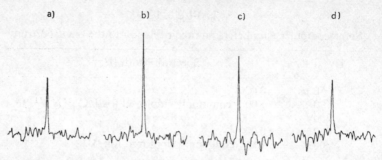

a)   b)   c)   d)

FIG. 3.11. Intensity of the quaternary aromatic carbon in ethylbenzene as a function of the flip angle (multipulse experiment). (a) $\alpha = 4°$. (b) $\alpha = 10°$. (c) $\alpha = 56°$. (d) $\alpha = 72°$.

α, and thus calculate a suitable pulse length. A qualitative relationship between T_1 and α is shown in Table 3.2. Values for α_E derived from this table should be considered as qualitative estimates since (a) under certain conditions[17b] α_E may be too small and (b) T_1 values are usually measured in the absence of molecular oxygen, whereas in routine measurements oxygen is not excluded. It is a naive approach to set t_p in the morning, having first determined the optimum signal/noise ratio on a test sample after n pulses, and then assume that this selection is valid for all problems (e.g. variable temperature experiments should be accompanied by varying t_p since relaxation mechanisms have a temperature dependence). We have attempted to avoid extensive data listing, however, we feel that since the dissemination of T_1 values for most low sensitivity nuclei is not widespread, a listing for one commonly studied nucleus is not out of place. In Table 3.3 and Fig. 3.12 are shown ^{13}C T_1 values for a wide range of molecular types. It is our feeling

TABLE 3.2

Suggested† flip angles, $\alpha_E{}^b$, as a function of T_1 for a 0.82 s acquisition time (5000 Hz spectral width, 8k FID)

$T_1(s)$	T_{cycle}/T_1	flip angle, α_E
0.2	4.1	90
0.5	1.6	70–75
1.0	0.8	65–70
2.0	0.4	45–50
4.0	0.2	30–40
8.0	0.1	20–30
50.0	0.016	3–4

† This is the so-called Ernst angle (see ref. 16), $\cos \alpha_E = \exp(-(t_{acq}+t_{post}))/T_1$

TABLE 3.3

Some representative ^{13}C T_1 values

Straight chains

	C_1 — C_2 — C_3 — C_4 — C_5 — C_6 — C_7 — C_8 — C_9 — C_{10}	Ref.
hexane	21·2 14·8 15·9	19
heptane	10·9 13·2 12·8 12·0	20
octane	12·8 10·8 10·1 9·6	19
decane	8·7 6·6 5·7 5·0 4·4	19

	Br-CH_{2_1} — C_2 — C_3 — C_4 — C_5 — C_6 — C_7 — C_8 — C_9 — C_{10}	
butyl	11·6 14·8 16·8 20·0[a]	21
pentyl	8·0 9·0 8·8 10·0 12·8[a]	21
hexyl	6·6 6·6 6·7 7·4 8·1 10·4[a]	21
heptyl	4·7 4·5 5·0 4·8 4·6 6·0 8·5[a]	21
octyl	3·6 3·8 3·1 3·6 3·6 3·9 4·6 7·6[a]	21
decyl	2·5 2·8 2·6 1·8 1·9 1·9 2·7 2·4 2·9 6·0[a]	21

R	R_1 — C_2 — C_3 — C_4 — C_5 — C_6 — C_7 — C_8 — C_9 — C_{10}	
$CH_2 OH$	0·6 0·7 0·7 0·8 0·8 0·8 1·1 1·2 1·6 3·0	22
$CO_2 H$	0·4 0·6 0·8 0·8 0·8 1·2 1·4 1·9 3·0	22
$CO_2 CH_3$	5·3 2·6 2·5 2·2 2·2 2·2 2·2 3·6 3·9 5·3	22
$CH_2 Ph$	1·4 1·2 1·1 1·1 1·1 1·3 1·3 1·8 2·8 3·0	22
$CH_2 I$	2·4 2·2 1·9 1·9 2·0 2·0 2·1 2·7 3·6 3·9	22
$CH_2^+ NMe_3$	4·4 5·0 5·2 6·3 8·6 14·3 N $CH_3 = 6·0$	23
$CH_2 NH_2$	13·4 13·4 15·0 12·1	27

"Small" molecules

$CHCl_3$	32·4	24
$CHBr_3$	1·7	24
$CH_3 OH$	17·5	24
$CH_3 I$	13·4	24
$CH_3 Br$	8·8	24
$\underline{C}H_3 CN$	13·1	28
$(CH_3)_2 \underline{C} = O$	36·1	24
$CH_3 \underline{C}O_2 H$	41·1	24
$C_6 H_5 \underline{C}OCH_3$	34·0	25
$C_6 H_5 \underline{C}OCl$	49·0	

[a] Observed $T_1 \times 3/2$

TABLE 3.3—*continued*

Substituted benzenes

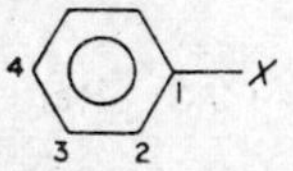

4⟨benzene ring⟩—X	C_1	C_2	C_3	C_4	C_α	C_β	Ref.
X							
H	29·3						
CH_3	58·0	20·0	21·0	15·0	16·3		26a
$CH_2 CH_3$	36·0	18·0	18·0	13·0	13·0	9·0	26b
$CH = CH_2$	75·0	14·8	13·5	11·9	17·0	7·8	26b
$C_6 H_5$	61·0	5·9	5·9	3·2			
$C \equiv C - H$	107·0	14·0	14·0	8·2	132·0	9·3	26a
OH	21·5	4·4	3·9	2·4			26a
NO_2	56·0	6·9	6·9	4·9			26a

Saturated Cyclic molecules $(CH_2)_n$ CHR

	C_1	C_2	C_3	C_4	C_5	
R = OH						
$n = 4$	2·5	4·1	3·3			29
$n = 5$	0·7	0·6	0·6	0·3		29
$n = 7$	0·3	0·3	0·2	0·4	0·2	29
R = OCH$_3$						
$n = 4$	23·4	17·3	19·4			29
$n = 5$	13·4	8·2	8·4	7·4		29
$n = 7$	7·7	4·6	5·1	5·0	4·4	29
R = H						
$n = 5$	17·0					30

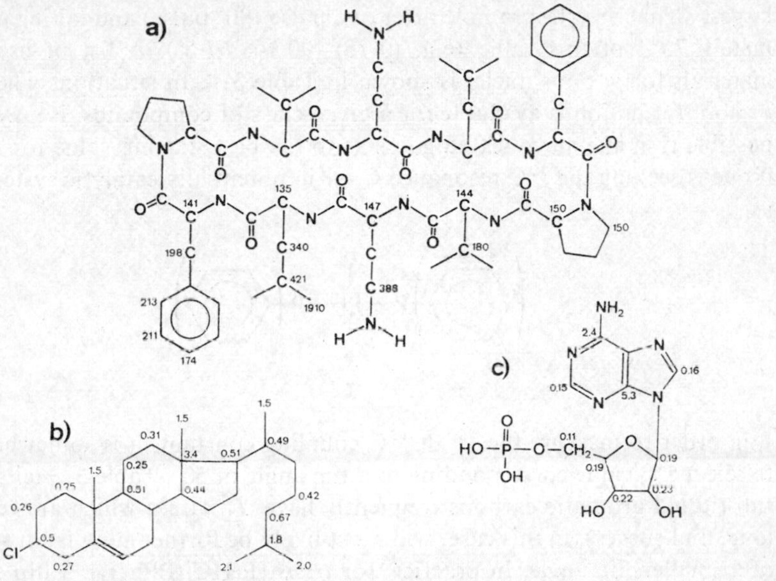

FIG. 3.12. T_1 values for (a) gramacidin-S (in ms),[31] (b) adenosine 5'-monophosphate[32] and (c) cholesteryl chloride[32].

that a knowledge of this parameter should and will shortly become as important as the chemical shift and coupling parameters.

A simple example of how one might extract, from Tables 3.2 and 3.3, information with regard to setting t_p, goes as follows. Assume we are interested in measuring the ^{13}C nmr spectrum of a steroid. Reference to Fig. 3.12 shows us that ^{13}C relaxation times, T_1, for the backbone carbons in such molecules are sometimes of the order of 0·2–1·0 seconds. For a 5000 Hz spectrum and 8 k FID, the flip angle to start with (Table 3.2) should be of the order of 65–70° (to begin with it would be best to see all the signals). This flip angle is estimated by finding that pulse length, t_p, which gives the

TABLE 3.4

Some useful test samples

Nucleus	Test substance	Lock substance
^{1}H	Ethyl benzene (1%)	C_6D_6
^{13}C	Ethyl benzene (90%)	C_6D_6 (10%)
^{15}N	Ammonium-chloride-^{15}N[a]	D_2O/H_2O (10%/90°)
^{31}P	Trimethyl phosphite (90%)	C_6D_6 (10%)

[a] > than 95 atom % enriched. The solution should be acidified to reduce the rate of the proton exchange

largest signal on a test sample after one pulse (90° pulse) and taking approximately 75% of this value (e.g. $(0{\cdot}75)$ $(90°) = 67{\cdot}5°$). A list of useful test materials for various nuclei is shown in Table 3.4. In situations where little or no information is available for a given class of compounds, it may still be possible to make an educated guess as to the best starting value for t_p. Thus if one is seeking the ^{13}C resonance C−1 in potentially catalytic systems such as

I

I, in order to measure the ^{195}Pt ^{13}C coupling constant, it is somewhat futile to select a t_p value corresponding to a flip angle of 50°. Table 3.3 tells us that substituted aromatic carbons frequently have T_1 values which are relatively long, and success, in this case, will most likely be forthcoming from selection of a smaller flip angle. In practice, for *trans*-(PtHCl(PPh$_3$)$_2$), a flip angle of $\approx 20°$ gave a reasonable result.

For cases where the T_1 of a given carbon is quite long but the concentration of this atom relatively high the introduction of a post-delay (see Fig. 3.10) can be fruitful. (A post delay can be better than selecting a longer t_{acq} since no *noise* is accumulated during this period.) Some software packages allow the entry of the parameter t_{post} directly via the teletype. This variable simply provides additional time during which T_1 relaxation can occur before the next pulse.

Amongst apprentice operators there is sometimes a tendency to believe that "pulsing rapidly", i.e. short dwell times and wide spectral widths, permits the experiment to be performed more quickly. It is desirable to pulse as rapidly as possible subject, of course, to restrictions resulting from relaxation phenomena; however, where pulse lengths are long (e.g. low frequency nuclei and/or crossed coil mode), the power across the spectrum may be non-linear.[18] A frequency which is pulsed for a length of time, t_p, must contain a continuous band of frequencies between $\pm 1/t_p$. The intensity, I, at some frequency, v is represented by

$$I(v) = \sin vt_p/\pi vt_p.$$

The practical consequence of this relation is shown in Table 3.5 where the % attenuation of power at frequency v is given as a function of t_p. Thus with a 5000 Hz spectral width and a 50 μs pulse the power distribution is not uniform. This non-uniformity can result in the observation of lines with reduced intensity relative to other lines in the spectrum.

TABLE 3.5

% Attenuation of power at frequency v shown as a function of pulse length t_p

	v (frequency from carrier wave)			
t_p	1kc	5kc	10kc	25kc
10	0	0·4	1·6	10
25	0·1	2·5	10	53
50	0·4	10	36	83

Number of scans

This parameter is entered via the teletype, and its selection is, of course, dependent on sample concentration and, sometimes, T_1 values of the nuclei under consideration. In practice, more often than not, this command takes the form of an infinity instruction; that is, "accumulate data until instructed to stop". In this fashion accumulation continues until the operator judges sufficient S/N has been obtained.

Digitization rate

The function of the digitizer has been touched upon in Chapter 2. In the example shown in that section, the S/N improvement was calculated to be 16.

$$\text{Total signal} = 2^{11} . 2^8 = 2^{19},$$
$$\text{Total noise} = 2^{11} . 2^4 = 2^{15},$$
$$S/N = \frac{2^{19}}{2^{15}} = 2^4.$$

For very weak signals (dilute solutions) a greater improvement is necessary, e.g. more scans.

This is possible if the digitizer resolution can be reduced to 8 bits (2^7 signal from each pulse). After 2^{12} pulses the memory will overflow (20 bit word).†

$$\text{Total signal } 2^{19},$$
$$\text{Total noise } 2^{13},$$
$$\text{improvement } S/N = 2^6 = \sqrt{N_s}.$$

Many spectrometers have software routines which automatically regulate the digitizer, however, in older generation spectrometers, the responsibility still lies with the operator.

† In actuality more data can be accumulated; see reference included in the footnote in Section 2.1.

Pre-delay

In theory the transmitter is "on" only when the receiver is off. In practice after a short high power pulse there is usually some ringing remnant in the receiver coil when the data acquisition begins. This is commonly called pulse break-through. It is frequently observable in the FID as large "spikes" in the first few channels of data. In order to allow the ringing to dampen, a short delay, generally of the order of one-or-two *dwell times*, is introduced between the end of the pulse and the beginning of the acquisition of data. This pre-delay should not be too long since the signal is decaying during this period and data is being lost. This loss of data is especially critical for nuclei with short spin–spin relaxation times (e.g. quadrupolar nuclei) since most of the signal may fall in the first few channels. Despite this precaution the FID will some-times show pulse breakthrough. After transformation, the breakthrough will often manifest itself in the form of a rolling baseline in the spectrum.†

Filter

In the discussion on folded lines we saw that the intensity of digitally folded lines (lines of higher frequency than the Nyquist frequency) was reduced by introducing a filter near the input to the computer. The presence of such a filter also cuts down on the amount of high frequency noise in the spectrum. It is usually selected such that its width matches the spectral width.

Receiver amplifier(s)

These should be adjusted such that the input voltage fills, but does not exceed, the digitizer. A voltage greater than that capable of being handled will be "cut off" or truncated.

Begin the experiment

The pulse sequence is initiated and the data accumulated until such time as the operator deems sufficient.

Reproduce and store the FID

Where sufficient supplemental memory is available this storage operation should always be performed. Most software routines have provision for copy-

---

† Several data processing methods have been suggested to eliminate this problem (i.e. depositing " O " in the first few channels of the FID, applying a " trapezoidal window " in this same area, a left arithmetic shift); however, these only cure the symptoms not the disease

ing and transferring data. Once the FID is copied and stored, the operator may transform the FID and decide if the spectrum is satisfactory; when not, he recalls the FID and continues data accumulation. Additionally, should the operator have performed a mathematical operation on the FID which has produced an unsatisfactory result he may rectify the error.

Data manipulation

Often the FID will correspond to a spectrum which has all the information required. However, occasionally the operator will desire a minor increase in the S/N or a slight improvement in the spectral resolution. In these instances it is possible, as was shown in Chapter 2, to mathematically operate on the FID to achieve these goals. Often the chemist knows in advance whether lines are likely to be closely spaced or have weak intensities and he may function accordingly. Most often a sensitivity enhancement will be performed. We have found that a useful starting value for the required exponent, TC, is approximately $-2\cdot5$ ($LB \sim 1\,Hz$) when using our "recommended" conditions (see above) for measuring nuclei such as ^{31}P, ^{15}N .... It is useful to keep in mind that T_2 values for broad lines are relatively short, with the converse true for sharp lines. Thus suppression, and sometimes complete removal of lines can be achieved by manipulating the FID (e.g. where sensitivity is no problem, *erasure* of the early section of the FID will remove very broad lines from the spectrum; conversely the latter part of the FID may be removed to enhance sensitivity when measuring quadrupolar nuclei).

Fourier transform the FID and select the "real" solution.

Phase correct the spectrum, using the phase "handles" provided. Keep in mind that lines with strongly differing T_1 values may be out of phase (so might a folded line) relative to the remaining spectral lines. A typical result after Fourier transformation is shown in Fig. 3.13.

Plot the spectrum taking care not to introduce any unwanted line broadening due to the time constant (filter) of the spectrometer *recorder*.

Calculate and print out the various chemical parameters (δ, J, integral) using the software routine provided. It should be remembered that the intensities and integrals suggested by the plotting and calculation routines may not correspond exactly to the "theoretical" values due to differences in T_1 values and/or NOE's. Sometimes a resonance may occupy two channels (quirk of fate) whereas a signal corresponding to the same number of atoms will fall in one channel. These intensity problems may be minimized by (a) a judicious choice of sensitivity enhancement and/or (b) increasing the number of data points in the transformed spectrum, (c) allowing sufficient time for T_1

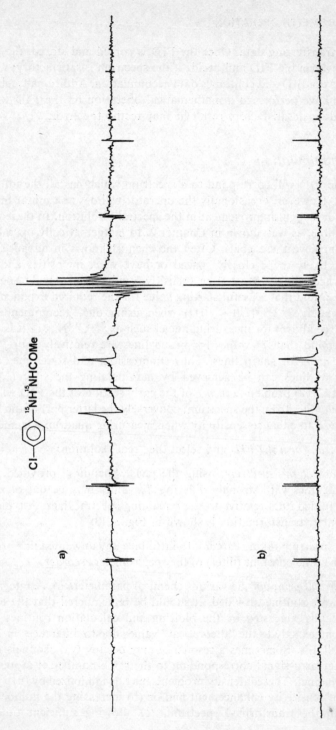

FIG. 3.13. (a) Uncorrected and (b) corrected ^{13}C frequency spectra of *p*-Cl-Ph-^{15}NH ^{15}NHCOCH$_3$ (DMSO-d$_6$ solution containing TMS).

relaxation, (d) the use of relaxation reagents (see Chapters 4 and 5) and (e) gated decoupling experiments.

For spectra where the signal-to-noise is poor it may not always be possible to quickly decide which threshold intensity to give to the computer. Too large a value results in its failure to recognize an important resonance whereas too small a value will result in the interpretation of noise as significant signals. Additionally, when resonances are very closely spaced the automatic calculation routine may not "resolve" them. In such cases the operator should utilize the manual cursor routines which allow the assignment of any individual data point to an appropriate frequency.

Review of suggested sequence for operation

1. Visually locate the lock signal using the oscilloscope.
2. Crudely shim the signal.
3. Reduce the width of the sweep, while alternately adjusting H_0 so that the lock signal is maintained on the oscilloscope.
4. Reduce the H_1 level for the lock nucleus to avoid saturation.
5. Lock the spectrometer (once the sweep is sufficiently reduced).
6. Shim on the lock signal for optimum resolution. Engage the automatic shimming control.
7. Select the mode of decoupling to be employed.
8. Select the parameters suitable for this mode (frequency, H_2 power).
9. Select the correct measuring frequency offset.
10. Select the spectral width.
11. Select the pulse length.
12. Select the rate of digitization (ADC).
13. Select the number of scans.
14. Select the size of the memory to contain the FID.
15. Select the post-delay (if necessary).
16. Select the pre-delay.
17. Select the correct filter for the given spectral width (if necessary).
18. Adjust the receiver amplifiers.
19. Begin the experiment.
20. Reproduce and store the FID (when supplemental memory is available).
21. Mathematically compensate for pulse breakthrough (if necessary).
22. Perform a sensitivity (resolution) enhancement operation.

23. Fourier transform the FID.

24. Phase correct the spectrum.

25. Read out (plot) the spectrum (real part).

26. Collect the data (δ, J) using the computer routines.

27. Leave the spectrometer in-lock.

3.4 Problems

1. In Fig. 3.14 is shown the ^{13}C spectrum of ^{15}N enriched p-chloro-aniline (^{15}N has nuclear spin I $= \frac{1}{2}$).

(a) Why are the intensities of the substituted carbons smaller?

(b) What non-chemical means might you use to remedy this?

2. In Fig. 3.15 the spectrum appears to be sitting on a "rolling baseline".

(a) How could this have occurred?

(b) How might it have been prevented?

3. Consider the theoretical spectrum shown in Fig. 3.16.

Given sufficient supplemental memory what could one do, mathematically, to favourably alter the appearance of the spectrum (other than increasing the spectral resolution and the number of scans).

4. Consider the aliphatic section of the ^{13}C spectrum (Fig. 3.17) showing the fragment $P-C_\alpha H_2-C_\beta H_2-C_\gamma H_2-C_\delta H_3$. Since it is recognized that $^1J(P, C) = 25$–35 Hz and $^3J(P, C) = 13$–16 Hz and that $^2J(P, C)$ and $^4J(P, C)$ are small and frequently not resolved, it would seem that the low field line of C_γ and the low field half of C_α are over-lapping.

Suggest ways of proving this.

5. In Fig. 3.18 is shown the ^{15}N spectrum of phenylhydrazine-$^{15}N_2$.

Suggest a reason why one set of lines appears as negative signals. (Hints: (a) the spectrum is not folded, (b) the gyromagnetic ratio of ^{15}N is negative.)

(Answers are located after the appendices.)

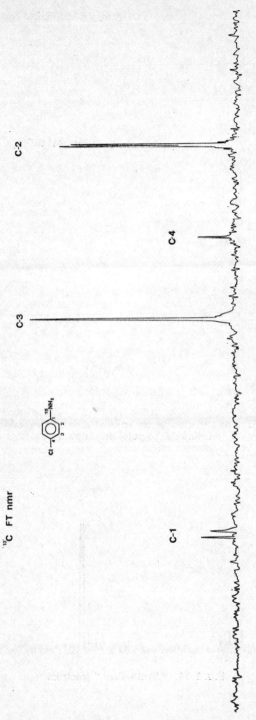

¹³C FT nmr

Fig. 3.14. ¹³C spectrum of *p*-chloro-aniline-¹⁵N.

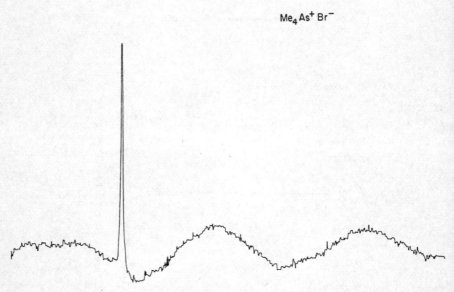

FIG. 3.15. ^{75}As FT nmr spectrum of tetramethyl arsonium bromide showing rolling baseline due to pulse breakthrough.

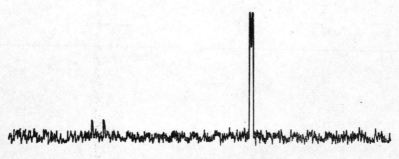

FIG. 3.16. " Synthesized " spectrum.

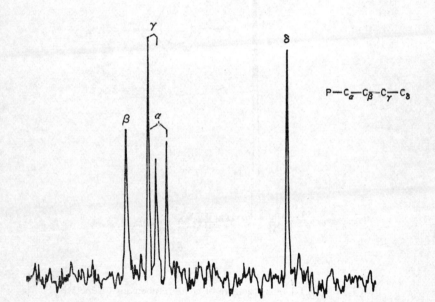

FIG. 3.17. ^{13}C spectrum (aliphatic carbons only) of *trans*-[PdCl$_2$(pyridine) (PBu$_3$^{n})].

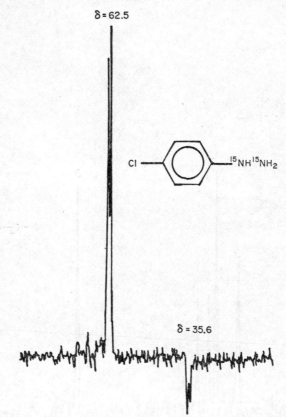

^{15}N FT nmr

$\delta = 62.5$

$\delta = 35.6$

FIG. 3.18. ^{15}N FT nmr spectrum of p-chloro-phenylhydrazine-^{15}N$_2$.

References

1. W. A. Anderson, *Phys. Rev.*, **102**, 151 (1956).
2. W. McFarlane, Nuclear magnetic double resonance spectroscopy, *in* "Determination of Organic Structures by Physical Methods" (Eds, F. Nachod and J. J. Zuckerman), Vol. 4.
3. W. V. Philipsborn, *Angew. Chem.*, **83**, 470 (1971); *Angew. Chem.* (Internat. Edn.), **10**, 472 (1971).
4. R. R. Ernst, *J. Chem. Phys.*, **45**, 3845 (1966); *Mol. Phys.*, **16**, 241 (1969).
5. W. A. Anderson and F. A. Nelson, *J. Chem. Phys.*, **39**, 183 (1963).
6. P. S. Pregosin, unpublished results.
7. J. H. Noggle and R. E. Schirmer, "The Nuclear Overhauser Effect, Chemical Applications." Academic Press, New York and London, 1971.

8. J. Solomon, *Phys. Rev.*, **99**, 559 (1955).
9a. K. F. Kuhlman and D. N. Grant, *J. Amer. Chem. Soc.*, **90**, 7355 (1968).
9b. G. Hawkes, W. Litchmann and E. W. Randall, submitted for publication, *J. Magn. Res.*
10. R. Freeman, *J. Chem. Phys.*, **53**, 457 (1970).
11. K. G. R. Pachler and P. L. Wessels, *J. Magn. Res.*, **12**, 337 (1973).
12. S. Sørensen, R. S. Hansen and H. J. Jakobson, *J. Magn. Res.*, **14**, 243 (1974).
13. S. A. Linde, H. J. Jakobson and B. J. Kimber, *J. Amer. Chem. Soc.*, **97**, 3219 (1975).
14. J. Feeney and P. Partington, *J.C.S. Chem. Commun.*, 1973, 611.
15. K. G. R. Pachler and P. L. Wessels, *J.C.S. Chem. Commun.*, 1974, 1038.
16. S. Schaeublin, A. Hoehener and R. R. Ernst, *J. Magn. Res.*, **13**, 196 (1974); P. Meakin and J. P. Jesson, *J. Magn. Res.*, **13**, 354 (1974).
17a. R. R. Ernst and W. A. Anderson, *Rev. Sci. Instr.*, **37**, 93 (1966).
17b. R. Freeman and H. D. W. Hill, *J. Magn. Res.*, **4**, 366 (1971).
18. P. Meakin and J. P. Jesson, *J. Magn. Res.*, **10**, 290 (1973).
19. N. J. M. Birdsall, A. G. Lee, Y. K. Levine, J. C. Metcalfe, P. Partington and G. C. K. Roberts, *J.C.S. Chem. Commun.*, 1973, 757.
20. J. R. Lyerla jr., H. M. McIntyre and D. A. Torchia, *Macromolecules*, **7**, 11 (1974).
21. Y. K. Levine, N. J. M. Birdsall, A. G. Lee, J. C. Metcalfe, P. Partington and G. C. K. Roberts, *J. Chem. Phys.*, **60**, 2890 (1974).
22. C. Chachaty, Y. Wolkowski, F. Piriou and G. Lukacs, *J.C.S. Chem. Commun.*, 1973, 951.
23. E. Williams, B. Sears, A. Allerhand and E. H. Cordes, *J. Amer. Chem. Soc.*, **95**, 4871 (1973).
24. J. C. Farrar, S. J. Druck, R. R. Shoup and E. W. Becker, *J. Amer. Chem. Soc.*, **94**, 699 (1972).
25. A. Olivson, E. Lippmaa and J. Past, *Eesti NSV. Tead. Akad. Toim Fuus-Mat.*, **16**, 390 (1967).
26a. G. C. Levy, J. D. Cargioli and F. A. L. Anet, *J. Amer. Chem. Soc.*, **95**, 1527 (1973).
26b. E. Breitmaier and W. Voelter, "^{13}C NMR Spectroscopy," p. 113. Verlag Chemie, Weinheim, 1974.
27. G. C. Levy, in *Accts. of Chem. Res.*, **6**, 161 (1973).
28. J. R. Lyerla jr., D. M. Grant and R. K. Harris, *J. Phys. Chem.*, **75**, 585 (1971).
29. G. C. Levy, R. A. Komoroski and R. E. Echols, *Org. Magn. Res.*, **7**, 172 (1972).
30. T. D. Alger and D. M. Grant, *J. Phys. Chem.*, **75**, 2539 (1971).
31. A. Allerhand and R. A. Komoroski, *J. Amer. Chem. Soc.*, **95**, 8228 (1973).
32. A. Allerhand, D. Doddrell and R. Komoroski, *J. Chem. Phys.*, **55**, 189 (1971).

4

Spin-lattice relaxation times

4.1 Mechanisms and significance

We have seen in the pulsed nuclear magnetic resonance experiment that the magnetization in the z direction is tipped away from the axis by some angle α. We have thought of this, pictorially (Fig. 4.1), as the absorption of energy which promotes a nuclear spin from the lower energy configuration (a), to the higher situation (b). We recognize that after some time, T_1, the situation represented in (a) will exist once again. We consider now the question of how this happens and the method most commonly used to measure the time necessary for this phenomenon to occur. It is obvious that the return from (b) to (a) is analogous to a magnetic resonance experiment and will require some form of an H_1 field with a component rotating at the Larmor frequency. This magnetic field, whose source is called the lattice, stems from the interaction of the nucleus with the neighbouring nuclei, solvent molecules or even other solute molecules via physical and chemical pathways. The suitable microscopic H_1 values may originate via the following mechanisms.[1]

1. Dipole–dipole.

2. Spin rotation.

3. Chemical shift anisotropy.

4. Scalar coupling.

5. Electric quadrupole interaction.

In the dipole–dipole term we are considering the magnetic field, H_{local}, experienced by a nuclear magnet X in the presence of a second nuclear magnet, Y. Since the two magnets are in motion in solution, this term depends on the frequency components of this motion.

The spin-rotation mechanism develops from fields generated by the motion of a molecular magnetic moment. This molecular moment stems from

FIG. 4.1. Nuclear energy levels.

the distribution of the electrons in the molecules. Rotation of electrons produces a current with which there is associated a magnetic moment.

Since the chemical shift screening tensor, σ, is usually anisotropic; i.e. on a relatively short time scale the components of σ; σ_x, σ_y and σ_z are not equal, a nucleus will see a fluctuating magnet field and the chemical shift interaction may provide a relaxation mechanism.

If a nucleus, X, chemically exchanges from site A to site B, where it is J-coupled to another spin, Y, then the local field at X will fluctuate as a function of the exchange rate (e.g. a sharp singlet when X is not bound to B, a multiplet when it is bound). The modulation of this coupling provides a relaxation mechanism commonly referred to as scalar relaxation of the first kind. When two nuclei are coupled, and the second of these has a T_1 value that is short compared with the inverse of the coupling constant, $1/J(X, Y)$, then the local field at the second nucleus fluctuates thus providing a second kind of scalar relaxation mechanism for this nucleus. Whether the change in the local field that modulates $J(X, Y)$ comes from chemical exchange or fast relaxation of one of the nuclei is immaterial from the point of view of the nucleus effected. This latter process is frequently the case when the first nucleus has nuclear spin $> 1/2$ since such nuclei possess an electric quadrupole. As this quadrupolar nucleus moves in solution, the quadrupolar coupling tensor becomes a function of time and provides a relaxation mechanism which for this type of nucleus is generally dominant.

Spin lattice relaxation times have been measured for many nuclei and at one time or another all of these mechanisms have been invoked to explain observed T_1 values. In the following paragraphs we will be concerned primarily with ^{13}C T_1 values since this parameter has been intensely investigated in recent times and because, in most cases,[†] ^{13}C T_1 values are dominated by the ^{13}C-H dipole–dipole interaction thus helping to simplify the interpretation of this parameter.

The necessity for these microscopic H_1 values to have the correct motional frequency in all of these relaxation mechanisms raises the question of molecular mobility. We express this in terms of the time, τ_c, that the molecule "remembers" one particular position. This parameter, τ_c, is commonly

[†] For small molecules such as CH_3OH, spin rotation is important. See T. C. Farrar S. J. Druck, R. R. Shoup and E. D. Becker, *J. Amer. Chem, Soc.*, 94, 699 (1972).

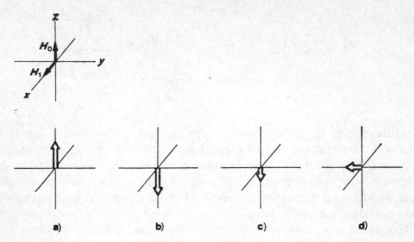

FIG. 4.2. The effect on the magnetization vector after each of the pulses of the inversion-recovery sequence.

referred to as the correlation time. A molecule which has only a short memory is moving rapidly in solution (τ_c perhaps 10^{-13}) while moderate organic molecules (e.g. steroids) move somewhat more slowly ($\tau_c \approx 10^{-10}$–10^{-11}).†

For a ^{13}C nucleus undergoing isotropic motion in the extreme narrowing condition ($\tau\omega_{13C} \ll 1$) the dipole–dipole relaxation term takes the form

$$1/T_1 = \frac{N\hbar^2 \gamma_C^2 \gamma_H^2}{r_{CH}^6} \tau_c \qquad (4.1)$$

where

N = number of hydrogens directly bound to carbon,

r_{CH} = carbon-hydrogen bond length,

γ = gyromagnetic ratio of the nucleus, and

τ_c = effective correlation time for rotational reorientation.

The implications from this simple relation are significant. Given a measured value for T_1, a reasonable value for the carbon-hydrogen bond length and a table of physical constants, we may obtain a qualitative estimate of *how the molecule is moving in solution*.‡ These results are subject to the usual experi-

† It is useful to recall that in any molecule there is a wide spectrum of motions but that we are interested in the density of one particular component. See T. C. Farrar and E. Becker, *in* "Pulse and Fourier Transform NMR", Chapter 4. Academic Press, 1971.

‡ Clearly, T_1 measurements will provide information within the realm of molecular dynamics. The separation of these sections is arbitrary.

mental and theoretical uncertainties (e.g. $\tau_c \omega$ is not always $\ll 1$ and molecular motion is not always isotropic), however, for a broad spectrum of organic compounds the ^{13}C T_1 experiment has proven quite useful.

Before proceeding into applications we should be conversant with the basic T_1 experiment and its problems.

4.2 Measurement

The inversion recovery method

The spin-lattice relaxation time, T_1, is most commonly measured using pulse sequences. A variety of sequences and modifications have been employed and evaluated.[2] Since it is our purpose to provide only an introduction to this area, interested readers are recommended to consult the supplementary references for details. The most commonly employed technique is the 180–τ–90 or inversion-recovery variation. In this method a 180° pulse inverts the magnetization from its positive equilibrium position, Fig. 4.2a, to that in Fig. 4.2b. After some time τ, during which the partial relaxation has occurred (Fig. 4.2c), a 90° pulse is applied. This rotates the magnetization vector into the y direction (positive or negative depending upon the relative length of τ) (Fig. 4.2d) and an FID may be recorded. If this experiment is repeated for various values of τ, making sure to allow sufficient time for complete re-equilibration of the magnetization after each sequence, spectra such as in Fig. 4.3 are obtained. For small values of τ the signal will be negative. For larger τ values the signal becomes positive and will eventually approach the intensity of a single 90° pulse when τ is greater than $3T_1$ values. Naturally, we may approximate T_1 by finding that τ value ($= T_1 \ln 2$) which produces an apparent nulling of the signal.

It is well known from the Bloch[3] equations that the decay of the magnetization in the z direction is

$$dM_z/dt = -(M_z - M_0)\, T_1 \tag{4.2}$$

where M_0 = magnetization at equilibrium before the pulse sequence, from which, by integration

$$M_z = M_0(1 - 2e^{-t/T_1}). \tag{4.3}$$

This may be expanded to

$$\left.\begin{array}{l} \ln(I_\infty - I_\tau) = \ln I_\infty - \tau/T_1 \\[6pt] I_\infty = \text{intensity for } \tau \gg T_1 \\[6pt] I_\tau = \text{measured intensity for a given } \tau \end{array}\right\} \tag{4.4}$$

^{15}N relaxation

(nat. abund.)

in H$_2$NCHO

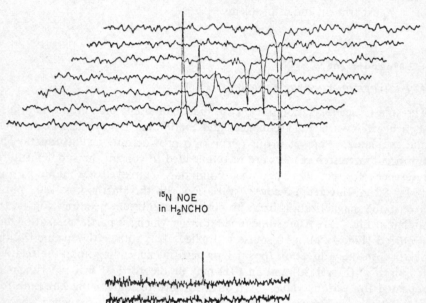

^{15}N NOE
in H$_2$NCHO

FIG. 4.3. ^{15}N Spin-lattice relaxation in H_2 ^{15}NCHO. The lower section of the trace shows an NOE experiment $\sim 3\cdot 93$ for ^{15}N. (Courtesy of Varian Associates, 1975.)

which will be recognized as an equation of the form $y = mx + b$, with $-1/T_1$ as the slope of the line. Thus a plot of the measurable quantity $\ln(I_\infty - I_\tau)$ v. τ will afford T_1.

The T_1 experiment is now automated in the most recent commercial spectrometers; however, there are a number of instances where problems may easily crop up. These have been discussed in some detail[4] and we list here some of the sources of error.

1. *An inhomogeneous* H_1 *across the sample.* To assist in this regard it is best to measure T_1 values using special cells which constrain the solution to the volume of the receiver coil.

2. *Misestimation of the 90 and 180° pulse lengths.*

3. *Data handling.* The cutting and weighing of peaks, although tedious, is thought to give accurate results.[4a] In this connection it is important to have a sufficient number of data points in the transformed spectrum to properly characterize the peak.

4. *The power of the pulse.* This is important if the null method ($T_1 = \tau/\ln 2$) is to be used to estimate T_1.

5. *Long* T_2 *values.* The presence of a residual signal in the direction of the receiver due to long T_2 values may also cause errors since these transverse components may be refocused by the proceeding RF-pulses (spin echo). Since the magnitude of the resulting components would be a function of the spin–spin-relaxation times, a determination of T_1 from a study of signal intensities would be incorrect. Methods have been described to disperse this transverse magnetization (e.g. by modification of the pulse sequence to include a short field-spoiling pulse before the 90° pulse).

6. *Failure to regulate the sample temperature.* The sample temperature should be regulated since relaxation mechanisms have a temperature dependence (e.g. T_1 in the dipole–dipole mechanism is directly proportional to temperature whereas the temperature dependence of T_1 in the spin-rotation mechanism is inverse).

7. *Solvent effects.* The solvent may provide both chemical and physical contributions to the values T_1. Thus a viscous solvent will affect the motional capability of the solute as may a hydrogen-bonding solvent.

8. *The presence of paramagnetic materials.* Paramagnetic substances, especially molecular oxygen, can significantly reduce T_1. Such impurities can be a significant problem for metal complexes where small amounts of paramagnetic impurities are formed, either from decomposition, equilibrium or other chemical processes.

Spin-spin-relaxation, T_2

While we will not review the advantages of spin–spin relaxation measurements in this section, it is useful, here, to record the sequence of pulses commonly used to measure this parameter.

We have seen that the signal in the receiver coil, the FID, decays as a

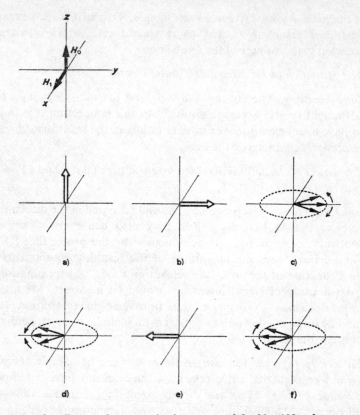

FIG. 4.4. The effect on the magnetization vector of the 90-τ-180 pulse sequence.

function of time. If the H_0 field were homogeneous then this FID would be a direct measure of T_2; however, this is not the case and the FID (and the width of the line observed in the nmr spectrum) is related to T_2^*, an effective spin–spin relaxation time. This may be represented as

$$1/T_2^* = 1/T_2 + (\gamma\Delta H_0/2).\tag{4.5}$$

One method of measuring T_2, using the so called spin-echo technique,[5] consists of applying the pulse sequence $90°-\tau-180°$ and observing an "echo" at time 2τ. This is diagrammatically shown, in the rotating frame of reference, in Fig. 4.4.

In Fig. 4.4b, after the first pulse, the magnetization has been tipped 90° into the y direction. Inhomogeneity in the H_0 field will cause some of the nuclear magnets to precede slightly slower than the rotating frame. These will appear to be moving counterclockwise as one views down the $+z$ direction while others will be somewhat faster and move clockwise (Fig.

4.4c). After some waiting time, τ, the 180° pulse inverts all the spins (Fig. 4.4d). The swifter nuclei continue to move clockwise, faster than the rotating frame, and, at time 2τ (Fig. 4.4e) the spins are rephased, affording an "echo". At times greater than 2τ dephasing is again observed. If no T_2 processes occurred, the amplitude of the echo signal would be equal to that following the 90° pulse. However, such processes *do* occur and thus the decrease in the amplitude of the echo is a function of τ, the time between the pulses. A plot of echo intensity v. τ will lead to T_2. This elegant, but simple, experiment has been modified in several fashions to overcome the problems of diffusion (Carr–Purcell[6]) and inaccuracy in H_1 (Meiboom–Gill[7]) but the principle remains unchanged.

4.3 T_1 applications

The inversion recovery experiment is not a new technique and neither is its use involving other than "routine" nuclei. Relatively new, however, is the wide-spread application of this technique to low sensitivity nuclei. In view of the increasing number of studies concerned with ^{13}C T_1 values[2] and the dominance in most cases of only one relaxation mechanism for this nucleus, we shall emphasize this type of study in this section. As shown by eqn (4.1), when the dipole–dipole mechanism dominates the ^{13}C relaxation one may relate T_1 to the molecular correlation time, τ_c,

$$\frac{1}{T_1} \propto N\tau_c/r^6.$$

At this juncture several points are worth noting.

1. If the molecule undergoes isotropic motion (it doesn't always) then the equation tells us that we will be able to differentiate CH, CH_2 and CH_3 carbon resonances, based on their T_1 values. This can be important when techniques such as proton off-resonance decoupling fail as a consequence of spectral complexity.

2. Since T_1 will depend on the inverse sixth power, $1/r^6$, of the carbon-hydrogen distance it will most often be the case that only protons one bond from the carbon will make a significant contribution.

3. τ_c may consist of more than one component. For simplicity it is usually represented as:
$$1/\tau_c = 1/\tau_{int} + 1/\tau_{over} \qquad (4.6)$$
where τ_{int} is the effective correlation time for internal motion of *a group* in a molecule and τ_{over} is the correlation time for overall reorientation of the molecule. Thus, if the molecule is large enough such that the over-

all motion is relatively slow (medium size organic molecules, polymers), $1/\tau_c$ is dominated by $1/\tau_{int}$ and we may obtain information on how each section of the molecule is moving in solution by measuring the ^{13}C T_1 values.

This is most elegantly demonstrated by the T_1 values (in seconds) for

$$\overset{0.65}{HO}-\overset{0.77}{CH_2}-\overset{0.77}{CH_2}-CH_2-\overset{0.84}{CH_2}-CH_2-\overset{1.1}{CH_2}-\overset{1.6}{CH_2}-\overset{2.2}{CH_2}-\overset{3.1}{CH_2}-CH_3$$

$$\overset{5.6}{H}-\overset{6.0}{CH_2}-\overset{5.2}{CH_2}-\overset{4.8}{CH_2}-\overset{4.3}{CH_2}-CH_2-CH_2-CH_2-CH_2-CH_2-CH_3$$

IIa

1-decanol[8] (IIa). The approximately 5-fold increase in T_1 values along the chain demonstrates that this molecule undergoes significant segmental motion with the $^{13}CH_3$ group moving much faster than the $^{13}CH_2OH$. The restriction in molecular motion as one approaches the hydroxyl group is believed to stem from hydrogen bonding phenomena. The T_1 values for the parent hydrocarbon show clearly that in the absence of a molecular "anchor", τ_c is considerably smaller.

The effect of hydrogen bonding on T_1 is also visible in some cyclic alcohols.[9] T_1 values for cyclopentanol are 2.5, 4.1, 3.3 s for carbons C_1 C_2 and C_3, respectively, whereas the corresponding values for methoxy cyclopentane are 23.4, 17.3 and 19.4 s.

Recently, the dependence of ^{13}C T_1 values in an entire series of decane derivatives $CH_3(CH_2)_8X$ (X = a carbon containing fragment) has been investigated.[10] Additionally, the basic relationship between τ_c and T_1 has been explored in some detail for n-alkanes[11] and n-alkylbromides[11] allowing (a) the calculation of diffusion coefficients for the molecule as a whole, about each C-C bond (in some cases) and (b) barriers to CH_3 rotation in linear and branched chain alkanes.

An interesting extension of this new knowledge of molecular motion in straight chains concerns ^{13}C T_1's in synthetic lecithins[12] and micellar

IIb

solutions. Thus in molecules such as IIb the ^{13}C T_1 values suggest that the structure is most tightly packed (and thus motion restricted) at the glycerol group (shortest T_1's) and that this area probably constitutes the main permeability barrier in such bilayers. The molecular motions of the fatty acid chains increase toward both the methyl group (0.1 to 3.3 s) and the choline head group (0.3 to 0.7 s). In aqueous solutions of the salts $[RN(CH_3)_3]^+ Br^-$ where $R = n$-hexyl, n-octyl, n-hexadecyl, the polar head group is itself sufficient to provide the "anchoring" which results in segmental motions along the chain [13]. While these motions are observable in the parent salt they are more pronounced in micelles of the same substances.

Once accepted as a relatively routine probe for molecular motions T_1 experiments may be used to assign ^{13}C spectra. In some instances the actual calculation of the T_1 values may not be necessary. In the assignment of the ^{13}C spectrum of the sugar derivative, stachyose (III)[14], the question of distinguishing the two galactose rings arose. The investigators noted that one of these is terminal *and* separated from the rest of the molecule by an $O-CH_2$ linkage and thus may have faster internal reorientation. The ^{13}C T_1

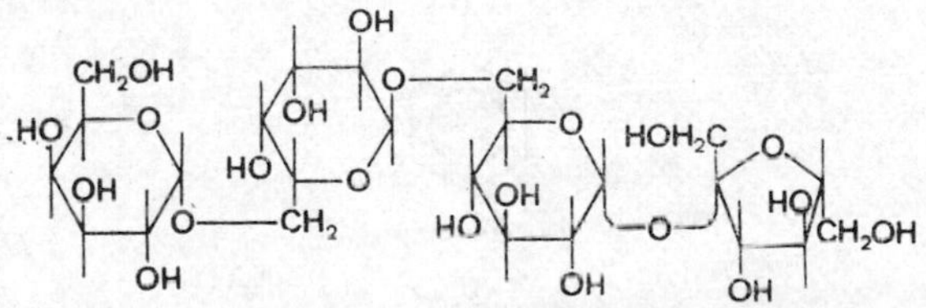

III

values for this carbohydrate reveal that one of the two galactose rings does indeed have longer carbon T_1 values than the other thus permitting a tentative assignment of the spectrum.

For the ^{13}C spectrum of mescaline (IV) it is possible to use the T_1 values

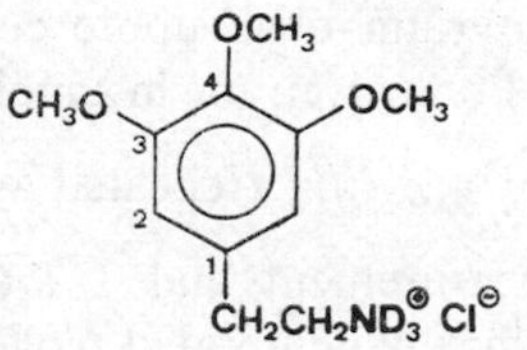

IV

for the *non-protonated* carbons to assist in the spectral assignment.[4c]. Since the relaxation of this molecule is dominated by the dipole–dipole interaction

F

the proximity to the benzene protons ($1/r^6$ term) is important. Thus C-1 within two bonds of both benzene and methylene protons has the shortest T_1 value whereas C-4 has the longest.

In assigning the ^{13}C spectrum of 3-bromobiphenyl, advantage has been taken of the fact that molecules may move anisotropically in solution.[15] Therefore, if rotation about a given molecular axis is preferred this may lead to modulation of the C-H dipole interaction (decrease in τ_c) and thus effect the magnitude of the T_1 values. The size of the increase in T_1, relative to the preferred axis, depends upon the term $\cos^2 \theta$, where θ = the angle between the preferred axis and the ^{13}C-H vector. Thus, in 3-bromobiphenyl (V), if the line joining 4 and 4' is the major axis, then C-4 and C-4' will have the shortest relaxation times since the C-H vector for these carbons makes an angle of 0° with the major axis. In a similar fashion we may continue the assignment.

$\underline{V}$

A novel approach to determining deuterium quadrupolar coupling constants via ^{13}C T_1 values has recently been described.[6] The ^{2}H relaxation time, $T_1(^2H)$ is dominated by the quadrupolar relaxation mechanism which is expressed as

$$1/T_1 \, (^2H) = 3(e^2 \, qQ/\hbar)^2 \, \tau_c/8 \text{ (extreme narrowing condition assumed),}$$

where $e^2 \, qQ/\hbar$ is the deuterium quadrupole coupling constant and τ_c the effective correlation time. For a given site in a molecule, if we know $1/T_1(^{13}C)$,

$$1/T_1 \, (^{13}C) = N\hbar^2 \, \gamma_c^2 \, \gamma_H^2 \, \tau_c/r^6 \text{ (repeated again for convenience),}$$

from a "normal" T_1 measurement and $1/T_1(^2H)$ from a deuterium T_1 measurement on the specifically deuterated molecular analog, then the value for the quadrupolar coupling constant may be calculated using eqn (4.7).

$$\frac{T_1(^{13}C)}{T_1(^2H)} = \frac{3r^6(e^2 \, qQ/\hbar)^2}{8N\hbar^2 \, \gamma_c^2 \, \gamma_H^2}. \tag{4.7}$$

Conversely, if the quadrupolar coupling constant is known then we may write

$$T_1(^{13}C)/T_1(^2H) = 19.9 N, \qquad (4.8)$$

where N = number of hydrogens bound to carbon, and we assume a value of r, the C-H bond distance, of 1.09 Å and a quadrupole coupling constant of 170 KHz. Thus it might be possible to predict ^{13}C or ^2H T_1 values given one of the relaxation times and the quadrupolar coupling constant. The utilization of ^2H "labelling" in ^{13}C nmr studies is increasing significantly and may lead more frequently to studies such as this one.

The ^{13}C T_1 study of cholosterol chloride (VI)[17] is interesting in that it represents an example of a molecule where both isotropic and segmental motion are observed. Within the experimental error the T_1 values of the

VI

protonated carbons on the ring backbone are inversely proportional to the number of bound protons (CH $T_1 = 0.52$ s. CH_2 $T_1 \approx 0.25$–0.27 s). Thus, the rotational motion of the backbone is nearly isotropic and it is possible to calculate the correlation time ($= 9 \times 10^{-11}$ s) using eqn (4.1); however, the D-ring side chain and the methyl groups show evidence for motions whose correlation times are less than that for the overall molecular motion. We may think of the D-ring as representing the anchor point with motion increasing in the side chain with increasing distance from the anchor. In discussing the T_1 results for this molecule, Allerhand and co-workers point out that when the molecular motion *is* isotropic, difficulties in interpreting the "off resonance" ^{13}C spectra may be circumvented with a knowledge of the ^{13}C T_1 values.

The application of ^{13}C T_1 studies is not restricted to relatively small molecules. Investigations of ribonuclease A and its oxidized derivative[18] as well as solid *cis* and *trans* polyisoprenes[19] have afforded information concerning segmental motions in such molecules. Although the interpretations are more qualitative due to the necessity of observing the response of many overlapping resonances, the technique provides relatively easy access to information which is not always readily forthcoming using other methods.

We have devoted our discussion primarily to ^{13}C; however, other low sensitivity nuclei, such as ^{2}H, are being increasingly studied. There has been a recent report[20] of the stereochemical dependence of ^{2}H T_1 values in a variety of organic molecules.

In particular, reasoning exactly analogous to that for the bromobiphenyl case has been employed to assign the deuterium spectrum of VII.[20] Since it

is known that the preferential axis of rotation can be envisioned by a line passing through the oxygen and nitrogen atoms, the *cis* CD$_3$ located perpendicular to the axis should have a longer ^{2}H T_1 value, whereas the *trans* CD$_3$ group (in analogy with a *para* carbon) located close to the major axis should have a shorter ^{2}H T_1 and indeed this is observed.

While not an "insensitive" nucleus, we also note a recent report concerning the configurational dependence of ^{1}H T_1 values in some carbohydrate derivatives.[21] In the glucopyranose derivative (VIII), the anomeric proton

shows smaller T_1 values (1.8 and 2.0 s for Cl$^-$ and Br$^-$, respectively) when disposed axially but larger T_1's (4.1 s for both halogens) when situated in an equatorial position. Thus we can anticipate that ^{1}H spin-lattice relaxation times may shortly be added to the long list of correlations between proton nmr and molecular structure. It would seem that the technique of measuring spin-lattice relaxation times has "graduated" from the physics to the chemistry laboratories.

References

1. A. Abragam, "The Principles of Nuclear Magnetism." Clarendon Press, Oxford, 1961.
2. See J. R. Lyerla Jr. and G. C. Levy, *in* "Topics in Carbon-13 NMR Spectroscopy" (Ed., G. C. Levy). Wiley Interscience, New York, 1974, Chapter 3.
3. F. Bloch, *Phys. Rev.*, **70**, 460 (1946); F. Bloch, W. W. Hansen and N. Packard, ibid., **70**, 474 (1946).

4a. I. M. Armitage, H. Huber, D. H. Live, H. Pearson and J. D. Roberts, *J. Magn. Res.*, **15**, 142 (1974).

4b. C. L. Wilkins, T. Brunner and D. J. Thoennes, *J. Magn. Res.*, **17**, 373 (1975).

4c. G. C. Levy and I. R. Peat, *J. Magn. Res.*, **18**, 500 (1975).

4d. P. Meakin and J. P. Jesson, *J. Magn. Res.*, **10**, 290 (1973).

5. E. L. Hahn, *Phys. Rev.*, **80**, 580 (1950).

6. H. Y. Carr and E. M. Purcell, *Phys. Rev.*, **94**, 630 (1954).

7. S. Meiboom and D. Gill, *Rev. Sci. Instrumentation*, **29**, 688 (1958).

8. D. Doddrell and A. Allerhand, *J. Amer. Chem. Soc.*, **93**, 1558 (1971).

9. G. C. Levy, R. A. Komoroski and R. E. Echols, *Org. Magn. Res.*, **7**, 172 (1975).

10. J. R. Lyerla Jr., H. M. McIntyre and D. A. Torchia, *Macromolecules*, **7**, 11 (1974).

11. Y. K. Levine, N. J. M. Birdsall, A. G. Lee, J. C. Metcalfe, P. Partington and G. C. K. Roberts, *J. Chem. Phys.*, **60**, 2890 (1974).

12. Y. K. Levine, N. J. M. Birdsall, A. G. Lee and J. C. Metcalfe, *Biochemistry*, **11**, 1416 (1972).

13. E. Williams, B. Sears, A. Allerhand and E. H. Cordes, *J. Amer. Chem. Soc.*, **95**, 4871 (1973).

14. A. Allerhand and D. Doddrell, *J. Amer. Chem. Soc.*, **93**, 2777 (1971).

15. G. C. Levy, *in Accts. Chem. Res.*, **6**, 161 (1973).

16. H. Saits, H. H. Mantsch and I. C. P. Smith, *J. Amer. Chem. Soc.*, **95**, 8453 (1973).

17. A. Allerhand, D. Doddrell and R. Komoroski, *J. Chem. Phys.*, **55**, 189 (1971).

18. V. Glushko, P. J. Lawson and F. R. N. Gurd, *J. Biol. Chem.*, **247**, 3176 (1972).

19. J. Schaefer, *Macromolecules*, **5**, 427 (1972).

20. H. Mantsch, H. Saitô, I. C. Leitch and I. C. P. Smith, *J. Amer. Chem. Soc.*, **96**, 256 (1974).

21. C. W. M. Grant, L. D. Hall and C. M. Preston, *J. Amer. Chem. Soc.*, **95**, 7744 (1973).

5

Commonly occurring non-routine problems

Although the study of "other" nuclei using FT techniques is advantageous there are some few problems with this method which are not normally encountered in CW ^{1}H spectroscopy. In addition to coping with the compromises inherent in the multipulse experiment (Chapters 1 and 4), we must be concerned with the concepts of data handling and the general problems of spin-lattice and spin–spin relaxation phenomena. Specifically, there are a few difficult situations which "pop-up" relatively often and we address ourselves, here, to how these are most commonly handled.

5.1 The suppression of solvent resonances

Measuring spectra in the FT-mode can sometimes prove difficult when one is faced with the problem of detecting weak solute signals in the presence of a large solvent signal. For example, in biological studies one often measures ^{1}H-FT spectra of dilute aqueous solutions in which the largest signal after each pulse is certainly due to the HOD-resonance. Not only do we encounter difficulties in the resulting spectrum due to the occurrence of beat frequencies and/or phasing problems but we must prevent the data destruction which will result if we exceed the vertical memory. Overflow in the computer memory must be avoided since this truncation process affects all frequency points in the transformed spectrum. In the CW-analog, CATing, overflow is not a serious problem since only one frequency is out of range. One approach to this type of dynamic range problem involves the accumulation and transformation of a limited number of FID's and subsequent addition of these data blocks in the frequency domain.[1] This solution is commonly called "block averaging".

Another serious problem arises from the limited word length of the analog-to-digital converter. The dynamic range of the digitizer must be large enough to allow the detection of a signal buried in the noise while at the same time sampling the largest signal of the spectrum without distortion.[2] It has been shown[3] that such a weak signal can be completely recovered if the quanta of digitization are smaller by ≈ 2 than the rms noise. If, as is the case in the presence of a very large solvent signal, the response of a single frequency covers a large dynamic range, the receiver gain of the spectrometer must be lowered in order to permit the signal to pass undistorted through the digitizer. In extreme situations the noise level can fall below the detection limit of the ADC. Then, inspite of signal averaging, the weak signal does not grow out of the noise and the researcher is thwarted. This situation requires that we somehow surpress the solvent signal prior to digitization.

One solution to the general problem involves selective irradiation of only the solvent resonance.[1] This type of double resonance experiment requires that we carefully adjust the decoupler power and frequency such that we do not disturb the rest of the spectrum. Practically, this may be achieved by inspecting the FID due to the solvent signal, on the scope and adjusting the decoupler frequency for minimum response.

A second commonly applied solvent suppression technique in ^1H spectroscopy is based on the large differences in ^1H-spin-lattice relaxation times between the solute protons ($< 1.0\,s$) and that of the HOD in D_2O solution ($5-15\,s$). The application of a pulse sequence similar to that applied in the determination of spin-lattice relaxation times by the inversion recovery technique ($= (180-\tau-90-T)_n$) gives impressive results if the interval between the pulses is selected such that the observing $90°$ pulse is applied just at that time when the "solvent magnetization" passes through zero ($= T_{1(HDO)}\ln 2$).[4] In this situation the solvent line essentially vanishes from the frequency spectrum; however, since the magnetization due to the solute will be almost completely recovered none of these resonances will be surpressed in the transformed spectrum. With this technique we solve both overflow and digitization problems.

The example of solvent signal elimination, given in Fig. 5.1, results from the observation and transformation of one transient. Clearly, since $T_{1(HDO)}$ is relatively long, multipulse† experiments of this type can be rather time consuming. An additional problem in this multipulse T_1 experiment concerns T_2^*, the effective spin–spin relaxation time. Normally, data are collected in the time t_{acq} which is frequently long enough such that the transverse magnetization has decayed to zero. Occasionally this will not be true and in the next time t_{acq} the FID will contain information from the previous $90°$ pulse. If not corrected this can lead to "echos" (see Chapter 4) in the final spectrum;

---

† One can calculate, in the steady state regime, the time interval, τ, for a given delay T, and solvent spin lattice relaxation time $T_{1\ solv}$ such that the magnetization of the solvent nuclei immediately before the $90°$ pulse is nulled.[5]

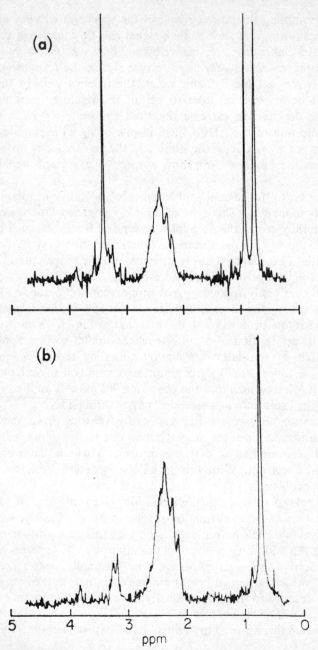

FIG. 5.1. Solvent (HDO) elimination in an ^{1}H spectrum using multipulse techniques, (a) with the HDO signal and (b) employing multipulse techniques. The co-solvent (acetone) signal is virtually eliminated as well.[4]

however, a modification of our original sequence to include a homogeneity "spoiling" pulse, e.g.

$$(180-HSP-\tau-90-T)_n,$$

solves the problem. In this sequence HSP is a short field gradient pulse (perhaps applied to a shim coil) which causes decay of the field homogeneity and thus rapid dispersal of any residual transverse magnetization.[5]

5.2 Shiftless relaxation reagents

When an nmr sample is doped with a paramagnetic compound the dipole–dipole interaction between the free electron and the nucleus can dominate the nuclear relaxation processes.[6] That the electron-nuclear relaxation is effective even at low additive concentrations and at relatively large distances between these dipoles is reasonable if we remember that $\gamma_e^2 \gamma_x^2$ (see eqn 4.1), where γ_e is the gyromagnetic ratio of the electron, is considerably larger than $\gamma_H^2 \gamma_x^2$. Common paramagnetic reagents are molecular oxygen, and complexes such as cobalt (II)-acetate,[7] chromium (III)-acetylacetonate[7-9] and iron (III)-acetylacetonate[7-9]. The latter are most frequently employed since they (a) are soluble in organic solvents, (b) do not chemically interact with the sample and (c) cause little or no change in the chemical shift of the species to be measured and only minor line broadening at the concentrations normally used (~ 0.1 M). Point (b) develops as a consequence of the nature of the bidentate ligand and the octahedral geometry of the complex.

Since the intranuclear C-II dipole–dipole relaxation mechanism is no longer significant, the importance of (or difficulties introduced by) nuclear Overhauser-effects is greatly reduced. Furthermore, due to the accelerated relaxation (γ_e^2 is much bigger, therefore $1/T_1$ increases) the sensitivity of an experiment involving slowly relaxing nuclei can be enhanced. For example the ^{13}C spin lattice relaxation time, T_1, for benzene has been measured at 23 s (viscosity 0.556 cP),[9] but in doped solutions in which the molarity of Cr (acac)$_3$ is 4.38×10^{-3}, 1.75×10^{-2} and 6.0×10^{-2} M, T_1 is shortened to 9.0, 3.05 and 0.98 s, respectively. Consequently, it is frequently the case that FT nmr experiments which require a meaningful comparison of signal intensities (e.g. quantitative analysis—see Chapter 6) are performed in the presence of such additives. Even more important, these reagents allow the *measurement* of certain types of carbon or nitrogen which, in their absence, have such long relaxation time that obtaining spectra has proved impossible.

A comparison of the effects of various paramagnetic additives can be performed by means of the relationship[6]

$$\frac{1}{T_1^e} = \frac{1}{T_1^{param}} - \frac{1}{T_1},$$

where T_1^{param} and T_1 are the spin lattice relaxation times measured with and without doping, respectively, and T_1^e is the dipole–dipole electron nuclear relaxation time. An equation has been derived by Abragam to calculate $1/T_1^e$, for equally sized hard-shell spheres.[10] This relation contains no explicit distant-dependent term but involves, for a given nucleus and reagent, the concentration, solution viscosity and temperature as the only variables. For ^{13}C-measurements with Cr(III) (acac)$_3$ as the additive, this parameter has a magnitude of the order of 30.[6] However, when specific intermolecular interactions between substrate and additive are present (e.g. weak coordination) this parameter increases significantly.†

Recently shiftless relaxation reagents have been used to achieve "chemical surpression" of long range proton coupling to quaternary carbons in ^{13}C spectra.[11] Indeed, if we work with sufficiently high concentrations of the relaxation reagent (0.1M Fe(acac)$_3$, 0.5−1.0M solute), we can "suppress" almost all *resonances* except the fully substituted carbons and the deuterated solvent thus achieving a chemical variation of the low power broad band decoupling technique.[12]

5.3 Fourier transform difference spectroscopy

Quite commonly in nmr, we perturb the spin system with a decoupling frequency or a relaxation reagent and observe the difference between the "normal" and perturbed spectra. If the effect produced is a small one (e.g. a homonuclear Overhauser effect) then the comparison may be difficult to discern. An interesting two-step method of highlighting the net effect involves the subtraction of FID's in the computer and is commonly referred to as "Fourier difference spectroscopy". Let us assume that we are performing the classical ^{1}H{^{1}H} decoupling experiment. We might proceed as follows. The ^{1}H FID is recorded in the usual manner and stored in a separate memory block. A second experiment using the same number of accumulations, receiver settings, etc. in the presence of the decoupling field is then performed. The two FID's are then subtracted‡ (most programs come equipped with this simple routine) and the difference FID, Fourier transformed.

In the resulting difference spectrum the irradiated resonance reappears (it is present in the normal spectrum) together with those resonances to which

---

† This type of effect should not be confused with isotropic shifts induced by so-called "shift reagents".

‡ If the conventional spectrum has not been measured previously it is best first to copy the FID, and then transform and print one duplicate while retaining the original for subtraction.

it is coupled. All others will disappear.† It has recently been demonstrated[13] that a series of such experiments offers advantage in the interpretation of complex ^{1}H nmr spectra when the resonance to be monitored is obscured by other absorptions. We have already pointed out that the combination of difference spectroscopy with the technique of homo- or heteronuclear selective population transfer (SPT) provides an FT-analog of the CW-INDOR technique.[14–16]

Recording the difference between a normal ^{1}H spectrum and one broadened by the presence of a paramagnetic ion has been shown [17] to be useful in pinpointing coordination sites in proteins. Since the paramagnetic species may associate at a specific location in the molecule, the only peaks in the difference spectrum will be those which are affected by the paramagnetic ion, thereby defining those resonances near the binding site. An additional application of nmr difference spectroscopy, in which broad lines are involved, is the convolution difference method. This combines sensitivity enhancement and the subtraction technique to improve spectral resolution.[17] After copying the FID, a severely decaying exponential is applied to one of the duplicates. After transformation of both FID's, subtraction of the processed spectrum, which now contains only broad lines, from the untreated data results in a spectrum with improved resolution.

Finally, we briefly mention a completely different form of "Fourier difference spectroscopy", developed by Ernst,[3] the practical consequence of which is that the usual field frequency-lock will no longer be required. The technique requires the presence of a reference compound which gives a signal considerably stronger than the sample signals. A normal pulse sequence is applied, but the signals are now detected with a linear detector instead of a normal phase sensitive detector. The difference between the sample and reference responses are derived with the aid of a low pass filter, digitized, and averaged as usual. The reference signal can be mathematically eliminated by subtracting a suitable function and the result is then Fourier transformed in the normal way to give the frequency spectrum. Perhaps the HOD signal may yet be useful!

Having equipped ourselves with the fundamental principles necessary to operate a Fourier nmr spectrometer we can now proceed to employ this technique. First, however, we might well ask "what new benefits will we receive from this technique?" In the following chapter, we have selected several readily measurable nuclei and show how nuclear magnetic resonance studies of these allow the researcher to probe a variety of chemical problems· in a direct manner.

---

† This is not perfectly true since lines lying close to the irradiation frequency may exhibit an intensity decrease and a slight shift. In addition the effects of the second radiofrequency may obscure the line to be decoupled in the difference spectrum.

References

1. F. Wehrli, "Solvent Suppression Techniques in Pulsed NMR", Varian Application Note, NMR-74-1.
2. D. G. Gillies and D. Shaw, *in* "Annual Reports on NMR Spectroscopy" (Ed., E. F. Mooney). Academic Press London and New York, 1972, Vol. 5.
3. R. R. Ernst, *J. Mag. Res.*, **4**, 280 (1971).
4. S. L. Patt and B. D. Sykes, *J. Chem. Phys.*, **56**, 3182 (1972).
5. F. W. Benz, J. Feeney and G. C. K. Roberts, *J. Magn. Res.*, **8**, 114, (1972).
6. J. R. Lyerla, Jr. and G. C. Levy, *in* "Topics in Carbon-13 NMR Spectroscopy" (Ed., G. C. Levy). Wiley Interscience, New York, 1974, Vol. 1, p. 122.
7. G. N. La Mar, *J. Amer. Chem. Soc.*, **93**, 1040 (1971); R. Freeman, K. G. R. Pachler and G. N. La Mar, *J. Chem. Phys.*, **55**, 4586 (1971).
8. S. Barcza and N. Engstrom, *J. Amer. Chem. Soc.*, **94**, 1762 (1972).
9. G. C. Levy and J. D. Cargioli, *J. Magn. Res.*, **10**, 231 (1973).
10. A. Abragam, "The Principles of Nuclear Magnetism." Clarendon Press, Oxford, 1961, Chapter 9.
11. U. Sequin and A. J. Scott, *J.C.S. Chem. Commun.*, 1973, 1041.
12. J. H. Sadler, *J.C.S. Chem. Commun.*, 1974, 809.
13. G. Massiot, S. K. Kan, P. Gonord and C. Duret, *J. Amer. Chem. Soc.*, **97**, 3277 (1975).
14. J. Feeney and P. Partington, *J.C.S. Chem. Commun.*, 1973, 611.
15. A. A. Chalmers, K. G. R. Pachler and P. L. Wessels, *J. Magn. Res.*, **15**, 415 (1974).
16. T. Bundgaard and H. J. Jakobsen, *J. Magn. Res.*, **18**, 209 (1975).
17. J. D. Campbell, C. M. Dobson, R. J. P. Williams and A. V. Xavier, *J. Magn. Res.*, **11**, 172 (1973).

6

"New nuclei". A bonus in the solution of chemical problems

Undoubtedly, the major advantage gleaned by the chemist from FT nmr stems from the additional sensitivity provided by the technique. This sensitivity is translated directly into new information by permitting the ready observation of a wide variety of different nuclei. Clearly, the nucelus which has received the most attention in this regard has been ^{13}C. Before 1971 the contents of reports involving this nucleus were easily summarized in relatively short reviews. Currently, there are no less than at least four books[1-4] and a new series,[5] "Topics in Carbon-13 NMR," dedicated to this nucleus.

While receiving less fanfare, a number of other nuclei have attracted increasing attention. Perhaps foremost on this list are nitrogen-15, phosphorus-31, and metals, in general. The spectra of these nuclei, are frequently relatively simple, since the chemical shift ranges and sensitivity of δ to structural changes are large, while the number of these nuclei per molecule relatively small. Additionally, these nuclei are frequently involved at the reaction site directly (e.g. the nitrogen of a histidine is directly coordinated to iron in haemoglobin) and thus their study may obviate the necessity of drawing conclusions from data taken at more remote centres. In the study of reactions involving metals, the direct observation of the metal centre may prove to be the most viable approach to the problem of understanding the nature of the metal complex in solution.

In a number of cases the speed of the FT technique has permitted the detection of intermediate species in solution as well as the measurement of kinetic phenomena. Thus, where the CW method requires minutes to cover a wide spectral region, the FID containing the same information is collected in seconds. This method in combination with sophisticated software routines and supplemental memory systems, allows automated relatively "fast"† kinetics using the nmr.

† Obviously not as fast as "stopped flow" and other methods but a significant improvement. The data are processed "en mass" sequentially, after all the experiments are finished.

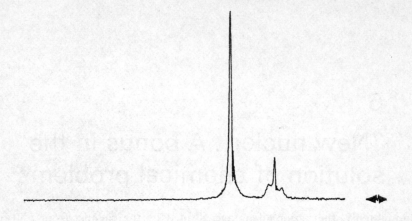

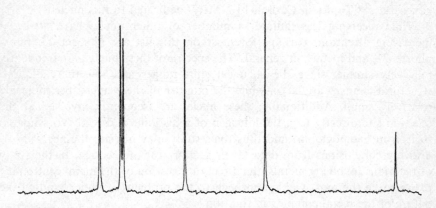

FIG. 6.1. The simplicity of ^{13}C nmr relative to ^{1}H nmr. The upper trace is the ^{1}H spectrum of *n*-decane; the lower trace the ^{13}C spectrum of the same material plus TMS.

In the following sections we have attempted to present a representative sampling of the types of problems whose solutions are readily forthcoming using "new" nuclei nmr. We have avoided extensive compilations, since these are available elsewhere, and hope instead that the chemist will derive from these few examples a feeling as to the potential for chemical research provided by these techniques.

6.1 Carbon-13. The advantage of spectral simplicity

The sensitivity problem encountered in measuring ^{13}C nmr spectra stems from the low natural abundance $(1\cdot1\%)$ of the ^{13}C-isotope $(I = \frac{1}{2})$† and the relatively small gyromagnetic ratio γ_c of this nucleus. Considering both factors leads to an intensity for this isotope of $1\cdot6 \times 10^{-4}$ relative to a proton at constant field. This low sensitivity is a problem common to most other nuclei with only a few exceptions (e.g. ^1H, ^{19}F). Furthermore, the relatively long spin–lattice relaxation times, T_1, of ^{13}C nuclei can present additional difficulties in the measurement of this nucleus. The total problem, however, must be contrasted with the enormous power of ^{13}C nmr as a tool for the elucidation of molecular structure. This develops directly from (a) the relatively large range of chemical shifts (>200 ppm) and (b) the spectral simplicity under the conditions of complete ^1H decoupling. Since C–C and C–H-scalar interactions are not detected, every magnetic site is represented specifically by an individual signal. In addition, a considerable amount of interesting chemistry occurs at carbon centres which do *not* bear protons (e.g. substituted aromatic centres) thereby relegating ^1H nmr to a "back seat" while elevating ^{13}C nmr.

An illustration of the relative simplicity of ^{13}C v. ^1H nmr is shown in Fig. 6.1. The ^1H-nmr spectra of simple paraffins exhibit only poorly resolved, frequently second order multiplets, whereas in the ^{13}C-nmr spectra usually all resonances can be resolved readily allowing spectral assignment‡ and integration (Fig. 6.1). Naturally, this aspect becomes even more important when we are concerned with compounds of high molecular weight (e.g. polymeric and biological systems).

The simplicity of the carbon spectrum can occasionally be used to advantage in an analytical sense. In fact, assuming that the appropriate precautions are taken to circumvent potential experimental problems (e.g. the influence of spin–lattice-relaxation times, nuclear Overhauser effects and an insufficient number of data points) the *quantitative* interpretation of ^{13}C-spectra offers possibilities which are comparable with those of other analytical methods (GC, uv, ir). For example, one problem in the analysis of fats and oils[6] involves the determination of the type and quantity of occurring fatty acids (usually consisting of a C-18 chain) in which there may be between zero and three double bonds. Normally, analysis is performed by methanolysis of the triglycerides and a subsequent gas chromatographic investigation of the

† It is worthwhile to remember that ^1H nmr profits considerably from this low natural abundance, since, if it were not so, we would probably have to record ^{13}C-decoupled ^1H-spectra.

‡ Those unfamiliar with the approach to assigning non-proton nmr spectra will find Appendix A helpful. In particular, a methodology for ^{13}C is detailed in Appendix B.

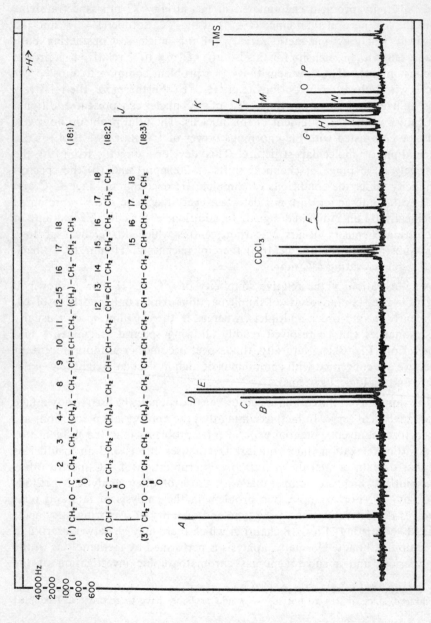

FIG. 6.2. ^{13}C nmr spectrum of linseed oil with 0.04 M[Cr(acac)$_3$] dissolved in CDCl$_3$[6]. A hypothetical triglyceride containing one, two and three olefin C-18 fragments. (Courtesy of Varian Associates. 1975.)

resulting mixture of esters. In the ^{13}C method the spectrum of the chemically intact oil can be recorded and one such example is shown in Fig. 6.2.

$$\overset{1}{CH_2}-O-\overset{2}{CO}-\overset{3-7}{CH_2}-(CH_2)_5-\overset{8}{CH_2}-\overset{9}{CH}=\overset{10}{CH}-\overset{11}{CH_2}-\overset{12}{CH_2}-\overset{13}{CH_2}-\overset{14}{CH_2}-\overset{15}{CH_2}-\overset{16}{CH_2}-\overset{17}{CH_2}-\overset{18}{CH_3}$$

$$\overset{}{CH}-O-CO-CH_2-(CH_2)_5-CH_2-\overset{12}{CH}=\overset{13}{CH}-\overset{14}{CH_2}-\overset{15}{CH}=\overset{16}{CH}-\overset{17}{CH_2}-\overset{18}{CH_2}-CH_2-CH_2-CH_3$$

$$\overset{}{CH_2}-O-CO-CH_2-(CH_2)_5-CH_2-CH=CH-CH_2-CH=CH-\overset{15}{CH_2}-\overset{16}{CH}=\overset{17}{CH}-\overset{18}{CH_2}-CH_3$$

IX

This spectrum was recorded from a sample of linseed oil known to contain four C-18 hydrocarbon fragments (three of which are shown attached to a glycerol fragment in IX) with 0, 1, 2 and 3 double bonds, respectively. Given, from independent measurements, that it is possible to find and assign several individual resonances (e.g. B, E and $O = C_{15}$, C_{16} and C_{17} of the 3 olefin fragment, $I = C_{16}$ of 2 the olefin fragment) as well as resonances which are a composite of a known carbon type (e.g. $P = C_{18}$ for all 4 fragments, $G = C_2$ for all 4) then it is possible to calculate the quantities of the individual components. In fact, there are more parameters than necessary and a least square analysis reduces the error. The results in this case (55·6% triene via ^{13}C nmr, 55·3% triene via GC) are quite reasonable and suggest that in some cases ^{13}C nmr can provide a non-destructive quantitative alternative.

When approaching a completely unknown structure the chemist will frequently consider, in the early stages, a determination of the molecular formula. In many cases this can be done by inspection of the mass spectrum. In several classes of compounds, however, insufficient volatility, instability or the occurrence of rapid fragmentation prevent the determination of the molecular weight from the M^+-peak. It has been demonstrated,[7] that simply counting the number of ^{13}C- or ^1H-signals in high-resolution-spectra provides an alternative technique. For this purpose ^{13}C-spectroscopy is superior to ^1H nmr, since the range of ^{13}C-chemical shifts is roughly 20 times larger than that of ^1H-shifts and because we can also *identify fully substituted carbons*. Naturally, the validity of this method is limited if complex signal-overlap occurs; however, perfect coalescence of signals is unlikely to be observed in view of the narrow ^{13}C linewidths (of the order of 1–2 Hz) and wide frequency shift range (≈ 5000 Hz). Still we sometimes encounter signals whose intensity and/or broadness suggests the overlap of signals. The unknown compound, whose carbon spectrum is shown in Fig. 6.3, revealed 27 well resolved ^{13}C resonances with one of these appearing somewhat broader and more intense than the others hinting at a 28 carbon entity. High resolution ^1H nmr suggested 38 protons and indeed there was a small mole-

G

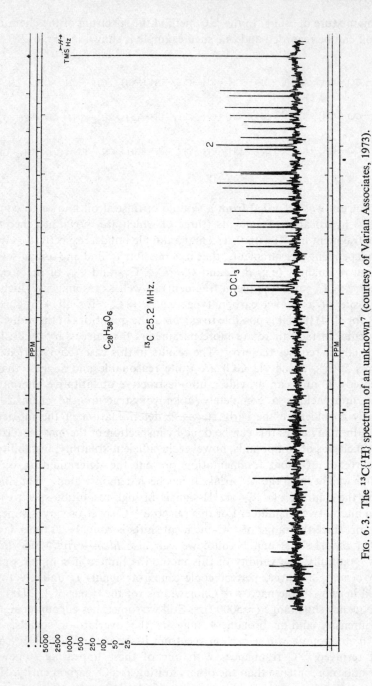

$C_{28}H_{38}O_6$

^{13}C 25.2 MHz.

CDCl$_3$

TMS Hz

2

PPM

PPM

5000 2500 1000 500 250 100 50 25

FIG. 6.3. The $^{13}C\{^1H\}$ spectrum of an unknown[7] (courtesy of Varian Associates, 1973).

cular ion peak in the mass spectrum at $470 \cdot 27100$ corresponding to $C_{28}H_{38}O_6$. We do not recommend ^{13}C nmr as a replacement for either mass spectroscopy or elemental analysis but rather that supplementary and occasionally unique analytical information can be forthcoming from its use.

The detection of functional groups, such as keto- or cyano-groups and especially fully substituted carbons, is easily performed by inspection of the ^{13}C spectrum. For example it is known from their spectrum that the allylic alcohol X on reaction with $Pb(OAc)_4$ incorporates an acetoxy group.[8] Further evidence concerning the structure of the reaction product, however, was not accessible from this technique. The ^1H-spectrum of this compound showed only high field aliphatic signals and a quartet signal at $\delta = 6 \cdot 1$ ppm all of which were consistent with either XI or XII. The ^{13}C-spectrum (see

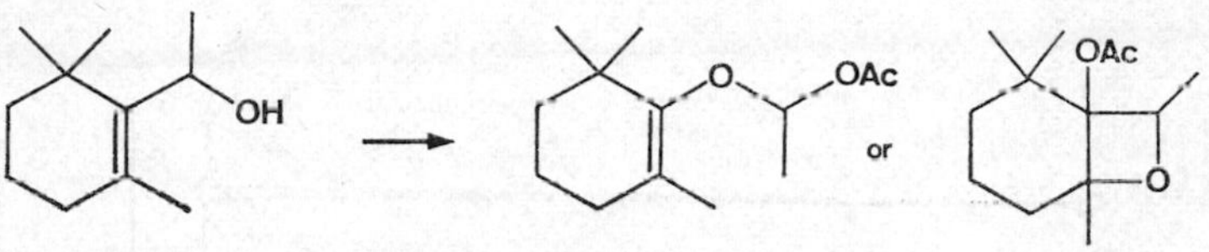

X   XI   XII

Fig. 6.4) allows an immediate decision in favour of structure XI since signals at low field for two olefinic carbons are readily observable. The aliphatic signals for XII would be expected to resonate at higher field by 30–40 ppm.

A similar approach has proven very useful in the study of valence isomerizations in which single or double bonds may be interconverted.[9] For those cases in which a change of hybridization occurs, simple inspection of the number of olefinic or aliphatic ^{13}C-signals allows a decision as to structure. An example of this type concerns the identification of bridged π-systems of

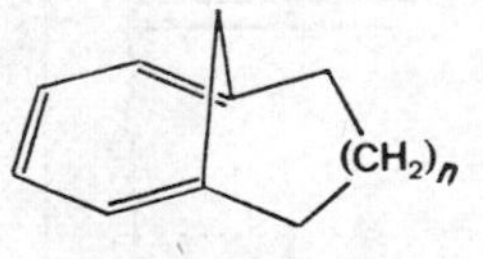

XIII a   XIII b

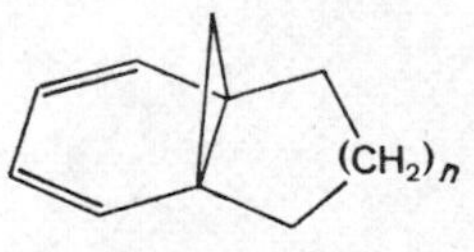

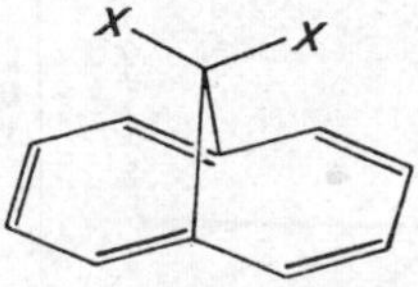

XIV a   XIV b

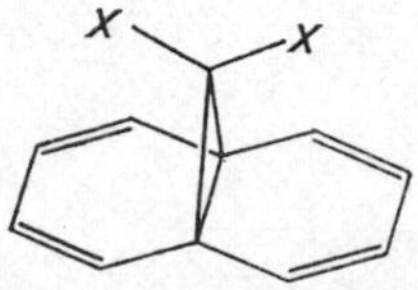

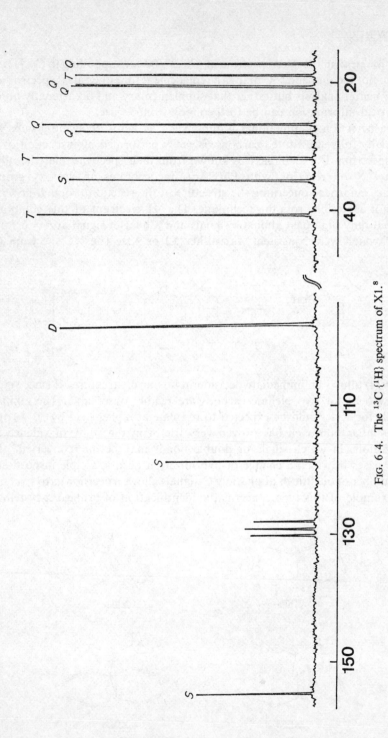

FIG. 6.4. The $^{13}C\{^1H\}$ spectrum of X1.[8]

the type XIII and XIV as cycloheptatriene or norcaradiene derivatives.[10, 11] It is clear that in these cases the question can be answered by counting the number of olefinic carbon signals (e.g. three for XIIIa, two for XIIIb). Since the bonding situation in XIIIa/XIIIb and XIVa/XIVb is variable and depends on the size of the alkylidene group $(CH2)_n$ and/or the nature of the CX_2-bridge, it is useful to have such a facile technique to distinguish them.

Another illustrative example concerns the investigation of the structure of the cyclooctatetraene-dimer.[12] It is clear from inspection of the hypothetical structures XV, XVI and XVII that one should expect quite different ^{13}C-spectra for those molecules. Since we observe four signals of equal intensity

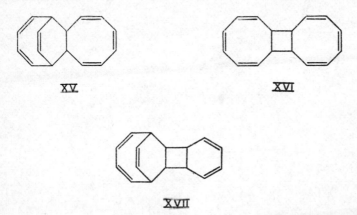

(three olefinic and one aliphatic) only structure XVI can be correct. This last example is obviously based on simple symmetry considerations. In many cases the ^{13}C-technique represents the ideal method for pointing out the existence of certain symmetry elements in rather complex molecules.

A basic question in the chemistry of cyclic unsaturated π-systems concerns the existence of localized or delocalized double bonds.[13] It is well known for $C_n H_n$-perimeter systems, the so-called annulenes, that cyclic conjugation will lead to stabilization in the case of $(4n+2)\pi$ electrons and to destabilization in the case of $4n\pi$ electrons.[14] The question of π-bond localization is also encountered in the bicyclic π-system heptalene.[15] It is obvious from Fig. 6.5 that the delocalized structure of heptalene, XVIII, assuming the molecule to be planar, possesses D_{2h}-symmetry, whereas the localized form XIX has only C_{2h}-symmetry. Accordingly, in the localized molecule only the magnetic sites 1 and 7, 2 and 8, ... 6 and 12 are equivalent, so that one would expect a six-line olefinic carbon spectrum with each resonance corresponding to two carbons. The delocalized species possesses two planes of symmetry through 1 and 7, and 4 and 10, respectively), and should give rise to a four-line spectrum (C-1, 7; C-2, 6; C-3, 5 and C-4) of relative intensity 1:2:2:1. Since

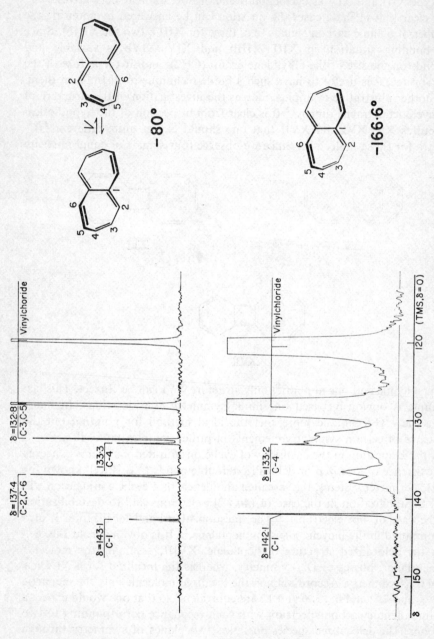

FIG. 6.5. The ^{13}C spectrum of heptalene in the fast and slow exchange situations The lower trace corresponds to heptalene in the localized structure and reveals that C_2 and C_6 as well as C_3 and C_5 are no longer equivalent.[15] (With permission of *Helv. Chim. Acta.*).

XVIII   XIX

at low temperature a six-line spectrum is observed it has been suggested that the lowering of the symmetry in heptalene stabilizes the π-system. This result is not influenced by the additional finding that the two localized iso-energetical structures undergo a rapid interconversion at elevated temperatures. The characterization of heptalene as a localized species using nmr is possible only by means of ^{13}C-spectroscopy, since the ^{1}H spectrum is not readily analysed at low temperatures. Nmr techniques which differentiate structure on the basis of molecular symmetry are also useful in transition metal chemistry.

The two tris(glycinato)-cobalt(III)-complexes, XX and XXI, have different colours and are thought to differ by the arrangement of the glycine fragments around the cobalt atom. Attempts to correlate colour (red and violet) with

XX   XXI

ligand configuration gave ambiguous results. The two complexes exhibit the expected ^{13}C resonances of the methylene and carbonyl carbons at high and low field, respectively;[16] however, the entire spectrum of the red complex consists of only two absorptions while that of the blue complex contains six lines. As can be seen from inspection of the formulae, the diastereomer with the facial configuration (XX) possesses a C_3-axis, whereas the diastereomer with the peripheral configuration has no symmetry relation between the fragments of the molecule. The ^{13}C-spectra reflect this difference and thus simply indicate that the red complex has a facial arrangement of the ligands. Completely analogous ^{13}C-studies have been performed for various diamagnetic metal complexes of cis- and trans-bis(ethylene diamine)cobalt(III).[17]

We conclude that, occasionally, a ^{13}C nmr spectrum, even *without* signal assignment, can be useful.

6.2 Correlation of ^{13}C chemical shifts with structural parameters

In this section we extend our basic considerations in order to establish correlations between δ^{13}C and structural properties. In a rather simple fashion we will consider the chemistry of unsaturated systems in terms of conjugation and charge distribution whereas for saturated compounds these correlations take the form of questions involving configuration and con-formation. Since ^{13}C nmr readily allows the observation of carbonyl groups, whose detection often involves only vibrational spectroscopy, we will begin with some comments in this area.

As is the case in ir spectroscopy, carbon nmr immediately distinguishes one form of carbonyl function from another. Thus, the ^{13}C resonance positions for the $>C = 0$ function, which generally appear in the range 160–220 ppm[1-4] can be readily subdivided into ketones (195–220 ppm), aldehydes (190–210 ppm) and acids (165–185 ppm). Within the range of ketones, we find that α, β-unsaturated carbonyl resonances tend to appear at the high field end. In addition, one of the olefin resonances of a system such as XXII is shifted considerably downfield from its normal position. Both of these observations

XXII   XXIII

are easily understood if it is accepted that a structure similar to XXIII makes a significant contribution to the total structure. In XXIII the carbonyl carbon has more olefinic character whereas the β-olefinic-carbon has a more cationic quality. This concept can be quite useful and has been applied[18, 19] in the differentiation of planar and twisted π-systems such as in XXIV–XXVI.

XXIV   XXV   XXVI

As a consequence of the *ortho*-methyl groups, torsion of the carbonyl with respect to the remaining π-system results in less conjugation with the benzene ring and thus in deshielding of the ketone carbon. Naturally, this alteration is also reflected in the position of the *para* carbon of the benzene ring where the shielding increases from XXIV to XXVI. This dependence of the chemical shift of olefinic carbons on π-conjugation readily explains

XXVII   XXVIII   XXIX

why C-1, C-3 and C-5 in XXVII–XXIX, the pentadienyl anion,[20] absorb at significantly higher field than do C-2 and C-4.

A particularly striking example of this type comes from a recent study of carbon-suboxide[21] (XXX) in which the ^{13}C resonance for C-1 was found 14·6 ppm *upfield* from TMS. If we recall that C-1 in allene appears at

$$O{=}C{=}C{=}C{=}O \qquad O{\equiv}C{-}C{=}C{=}O$$

XXX   XXXI

213·5 ppm *downfield* from TMS, this observation demands a consideration of a structure such as XXXI.

In MO models the bonding properties of π systems are commonly described by parameters such as π charge density or π bond order.[22] Since ^{13}C nmr seems responsive to charge on carbon, it is interesting to ask whether carbon shifts confirm MO predictions. For naphthalene (XXXII) and azulene (XXXIII) which are representative examples of alternant and non-alternant hydrocarbons, respectively, the ^{13}C-resonances of the alternant hydrocarbons cover a much smaller range (< 10 ppm) than those of the non-alternant ones (~ 20 ppm).[23-25] This finding is in keeping with the suggestion from MO theory that only for the former series is a uniform charge distribution to be expected.

XXXII   XXXIII

It becomes fairly obvious from these examples that, for a given class of compounds, $\delta^{13}C$/structure and $\delta^{13}C$/charge density correlations can often be derived in a straightforward manner. For cyclic and acyclic ketones there have been correlations of $\delta^{13}CO$ with ring size[26] and frequency of the $n \rightarrow \pi^*$ transition,[27] respectively. Similar investigations have led to a relationship between $\delta^{13}CO$ and half wave polarographic potentials in some quinoid structures.[28] There has been a report for benzocyclobutene derivatives of a correlation between reactivity and ^{13}C chemical shifts.[29]

Perhaps the classical example of a chemical shift-structure correlation concerns carbonium ions. The ^{13}C spectra of carbocations[30] reveal one line whose exceptionally low field position is a consequence of the localized positive charge on carbon. Thus, the cationic carbon of the tert-butylcation resonates at about 330 ppm (the total range of shifts is normally ~ 200 ppm). In view of the exceptionally low-field position of this charged centre, it has been possible to clarify certain classical problems involving structures such as XXXIV and XXXV. For the open situation we expect an average (bromo aliphatic and cation) ^{13}C resonance position for the two carbons of the order of > 100 ppm to low field of TMS, whereas the cyclic structure should approximate a charged heterocyclopropane and thus resonate at a higher field.[30]

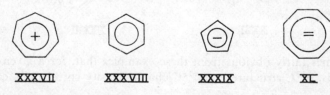

XXXIV   XXXV   XXXVI

In fact the carbon atoms of the ethylene bromium ion resonate at 75·6 ppm; a much higher field than one would expect for the average value of a primary alkyl carbonium ion, and a bromoaliphatic carbon, strongly supporting the literature in favour of the cyclic structure. In a similar fashion evidence from ^{13}C has been derived supporting the bridged structure XXXVI.[30]

From our knowledge of the ^{13}C position of carbocation centres we can estimate the efficiency of a neighbouring group in delocalizing the positive charge. Thus, in the investigation of a series of phenyl and cyclopropyl cations it has been concluded,[31] based on the chemical shift of the carbonium ion, that the phenyl ring is much more effective in withdrawing the

XXXVII   XXXVIII   XXXIX   XL

TABLE 6.1

^{13}C δ values in heptalene and its dianion[a]

	C_1	C_2	C_3	C_4	$\langle\delta\rangle$[b]
	113·9	111·1	91·7	103·8	103·7
	143·1	137·4	132·8	133·3	136·1

[a] Measured in ppm from THF and then corrected to TMS
[b] Centre of gravity of the ring carbon signals

positive charge than is the cyclopropyl ring. The cationic centre in XLI absorbs 26 ppm to higher field than does that of XLII. However, the conclusion that the phenyl group is energetically a "better" substituent is somewhat premature[32] since the measurement of the rates of solvolysis of both the corresponding tertiary-2-propyl-p-nitrobenzoate esters leads to the inverse conclusion.[33]

Not surprisingly, in various ionic systems we find a marked dependence of ^{13}C chemical shifts on charge densities. For the aromatic systems tropylium cation, benzene, cyclopentadienyl anion and cyclooctatetraene dianion (XXXVII–XL), a linear relationship between these parameters has been suggested.[34] In general a value of ≈ 160–200 ppm/electron is considered to be the correct order of magnitude.[35]

An additional case is demonstrated in Table 6.1 for heptalene and its corresponding dianion. The latter is noteworthy in that it represents a case of a delocalized 14-π charged system whose thermal stability exceeds that of the parent compound. Interestingly, the charge density 95, reflected by

the upfield shifts in the dianion, is not uniform. While the ^{13}C heptalene spectra are quite simple and give $\Delta\delta$ readily, the corresponding ^1H spectra suffer from both homonuclear coupling and the uncertainty in the magnitude of the ring current.

It is possible by the combined application of various assignment techniques to analyse the ^{13}C spectra of rather complex molecules,[36] and thus to extend

the ^{13}C-studies concerned with charge distribution and conjugation to biological systems. For example, it has been shown for porphyrin[37] that
significant resonance effects exist which are assumed to be due to the existence
of a preferred conjugation pathway in the "inner" 16 membered ring. By
comparing the ^{13}C spectra of "chlorophyll a" itself (XLIII) and its
magnesium-free derivative pheophytin it is concluded that magnesium-
coordination causes a significant downfield shift for carbon signals from
rings I and III and an upfield shift for those from rings II and IV. The
former finding indicates anion formation in rings I and III. This is thought to

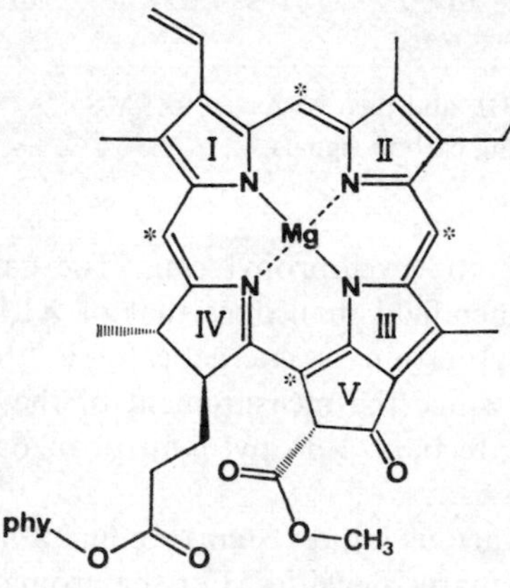

XLIII

occur via delocalization of the non-bonding electrons on nitrogen into
the π-system.[38]

Quite naturally ^{13}C nmr has become increasingly popular in the study of
organometallic compounds since carbon atoms directly bound to metals often
resonate at the extremes of the ^{13}C range. A particularly illustrative example
concerns species such as XLIV and XLV which are commonly thought of as
metal–carbene complexes.[39] The ^{13}C carbene resonances, C*, in molecules
of this type are found at extremely low field positions ($+250 \rightarrow 350$ ppm from

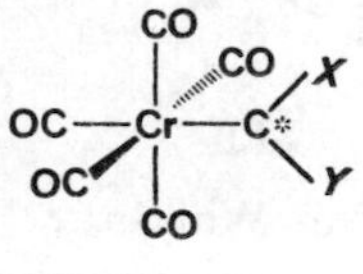

$X = OCH_3, NMe_2$
$Y = CH_3, Ph$

XLIV

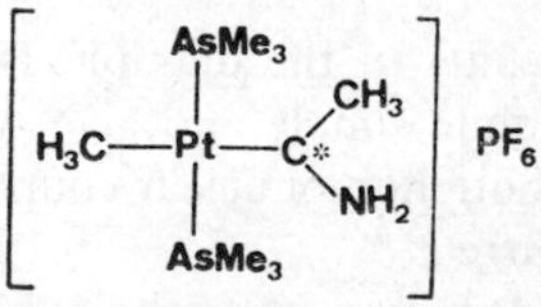

XLV

TMS) thus confirming their similarity to trialkyl carbocations. In fact the consideration of such carbene complexes as metal stabilized carbonium ions is based primarily on ^{13}C nmr studies. At the other extreme of the ^{13}C chemical shift range (7–31 ppm *upfield* from TMS) one finds the CH_3 resonances of the square planar platinum methyl complexes *trans*$[CH_3Pt(AsMe_3)_2L]PF_6$ (see XLV).[40] The ability to readily measure these carbon resonances provides a source of valuable information from both $^1J(^{195}Pt, ^{13}C)$ (^{195}Pt has $I = \frac{1}{2}$) and $\delta^{13}C$, with regard to the way in which the metal-carbon bond is formulated. Thus $^1J(^{195}Pt, ^{13}CH_3)$ may vary from $\approx$ 400–700 Hz as a function of L, the fourth ligand.

The carbon probe has also been used in an attempt to define the nature of the platinum–olefin bond. Thus, when mono-substituted olefins bind to

XLVI   XLVII

the metal, one can ask if both olefin carbons participate equally in the platinum–carbon bonds which are formed. A study of some platinum–styrene complexes containing the fragment XLVI, with Y varying from NO_2 to NMe_2, showed that both δC_1 and $^1J(^{195}Pt, ^{13}C_1)$ decreased as the electron donating ability of Y *decreased* while the screening of C_2 and $^1J(^{195}Pt, ^{13}C_2)$ *increased* as a function of this same parameter. The values $^1J(Pt, C_1)$ are always greater than those for $^1J(Pt, C_2)$. These data together with ir results have permitted the suggestion that an ionic form such as XLVII in which C_1 is covalently bound to Pt may make a significant contribution to the styrene–platinum bond.[41]

In addition to providing suggestions as to what nmr parameters we *can* use in the study of platinum–olefin complexes, carbon nmr has been able to indicate which ones might be misleading. In a recent investigation[42] of the complexes XLVIII and XLIX it has been shown that although the chemical

XLVIII   XLIX

shifts and coupling constants of the CH_3 groups can vary significantly (a factor of two in $^2J(^{195}Pt, ^{13}CH_3)$!), the values δCH and $^1J(^{195}Pt, ^{13}C)$ are only very slightly ($<5\%$) different. Since the one bond interactions are thought[42] to more directly assess the strength of the metal–olefin bond, the nmr parameters of the CH_3 group (^{13}C or ^1H) are *not* a good probe. It is clear, even from these few examples from transition metal chemistry, that the major constituent of the molecule need not be carbon for ^{13}C nmr to be a valuable tool.

The potential of ^{13}C nmr for determining molecular configuration is considerable since the ^{13}C aliphatic substituent effects (see Appendix A) are dependent upon configuration. This fact was exploited in the study[43] of some santonin stereoisomers of type L in which the ultimate objective was the determination for pseudosantonin, LI, of the configurations at C-8 and

I   II

C-11 as well as the geometry of the lactone fusion. The compounds L may possess *cis* or *trans* fused lactone rings and/or pseudo axial or pseudo equatorial α-methyl groups and thus there are four isomeric possibilities. Since early ^{13}C work on methyl cyclohexanes[44] and cyclohexanols[45] has shown that carbon chemical shifts depend on molecular configuration, the first step was the measurement and interpretation of the ^{13}C spectra of the four parent compounds in order to assess the validity of these relations in more complicated systems. Since it has been long recognized that the alteration of substituents (type *or* configuration, see Appendix A) significantly affects the resonance positions of all carbons within 3-bonds of the site of change,[46] it was not surprising to find several configurational dependences in the basic santonin system. e.g.

$C_6 \simeq$ 81 ppm when the lactone is fused *trans* ⎫

$C_6 \simeq$ 77 ppm when the lactone is fused *cis*   ⎬ Ring fusion

$C_9 \simeq$ 38–39 ppm when the lactone is fused *trans*

$C_9 \simeq$ 35 ppm when the lactone is fused *cis*   ⎭

$C_8 = 23\cdot3$ and $23\cdot4$ ppm when $-CH_3$ pseudo equatorial ⎫

$C_8 = 18\cdot3$ and $20\cdot3$ ppm when $-CH_3$ pseudo axial   ⎬ C_{13}

$C_{13} = 12\cdot5$ and $14\cdot9$ ppm when $-CH_3$ pseudo equatorial ⎨ configuration.

$C_{13} = \ \ 9\cdot6$ and $\ \ 9\cdot9$ ppm when $-CH_3$ pseudo axial ⎭

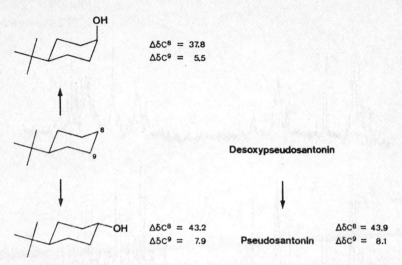

$\Delta\delta C^8 = 37.8$
$\Delta\delta C^9 = 5.5$

Desoxypseudosantonin

$\Delta\delta C^8 = 43.2$ Pseudosantonin $\Delta\delta C^8 = 43.9$
$\Delta\delta C^9 = 7.9$ $\Delta\delta C^9 = 8.1$

FIG. 6.6. Differences in ^{13}C δ values between pseudosantonin and desoxypseudosantonin[43] as compared with model systems. The numbers 8 and 9 refer to structure L. [44, 45]

All of these results are consistent with previously noted substituent effects. Thus, when pseudosantonin (LI) showed $C_6 = 77.5$ and $C_{13} = 15.6$ ppm and desoxypseudosantonin, the molecule without the C_8 hydroxyl, $C_6 = 76.6$ and $C_{13} = 15.2$ ppm, respectively, two of the three configurations (a *cis* lactone ring fusion and a pseudo equatorial C_{13}) seemed certain.† It remained now to decide whether the C-8 hydroxyl group was equatorial or axial. Since, as shown in Fig. 6.6, the effects of "substitution" on the carbon resonance positions of pseudosantonin seem to agree with those observed for an equatorial hydroxyl in cyclohexanols the third configuration is determined and the total objective is achieved. Perhaps the most interesting point of this example is that, after laying a proper basis, it required only one experiment to make *three* configurational assignments.

The use of model systems has also been shown to be valid in the conformational and configurational characterization of considerably more complex saturated molecules. Peptides and proteins are known to exist in linear arrays of amino acid fragments linked by peptide bonds. For a given sequence, however, a large number of conformations are possible due to rotations around single bonds. For small peptides, it has been demonstrated[47] that ^{13}C spectra can be interpreted by a straightforward analysis of the characteristic backbone-resonances. In some cases (e.g. the *cis* and *trans* isomers about the peptide bond in proline[47]) differentiation between specific conformations is possible as well. The information derived from these fundamental investi-

† A number of other resonance positions confirm this as well.

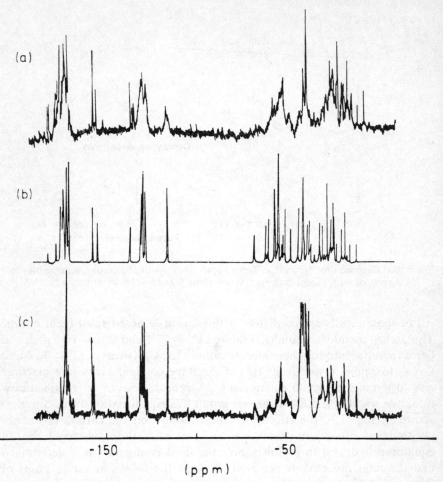

FIG. 6.7. Natural abundance $^{13}C\{^1H\}$ FT nmr spectra at 25·14 MHz of the basic pancreatic trypsin inhibitor. (a) Native protein in D_2O solution, (b) hypothetical spectrum computed from amino-acid spectra and (c) denatured protein in DMSO solution.[48]

gations has provided the basis for studies of higher proteins, which, since they consist of chains with sometimes several hundred amino acids, show ^{13}C spectra that are often only weakly resolved. The approach in such a complicated problem stems[48, 49] from the knowledge that, in the biologically functional "native" form, the amino acids occur in a unique arrangement in space the alteration of whose topography can be detected by characteristic changes in the ^{13}C chemical shifts. This concept leads directly to the type of experiment shown in Fig. 6.7 dealing with a "basic pancreatic trypsin inhibitor". In Fig. 6.7 we compare: (a) the ^{13}C-spectrum of the "native" protein in D_2O, (b) a hypothetical ^{13}C spectrum, simulated on the basis of the single amino acid resonances and (c) the ^{13}C spectrum of a denatured form of the protein in DMSO. The denatured form is believed to exist in a random coil conformation in which the amino and side chains are extended freely into the solvent. In this situation all the side chain groups appear as they would in the individual amino acids. This is confirmed by the similarity of spectra (b) and (c), both of which differ significantly from (a).

In studies concerned with the characterization of carbohydrates in solution it has been deduced from polarimetric measurements on D-fructose that a

LII   LIV

LIII   LV

considerable amount of β-D-fructofuranose is present at equilibrium in water and that the two α-anomers play a minor role.[50] This situation was easily clarified through the ^{13}C-nmr spectrum of D-fructose.[51] Since the ^{13}C resonance of the acetal carbon, C-2, has a configurational dependence and falls in a unique region of the spectrum it was possible to detect three distinct types of C-2 in the ratio $6 : 3 : 1$ corresponding to the species β-pyranose (LII), β-furanose (LIII) and α-furanose (LIV) (plus a trace of α-pyranose, LV). Thus, although nmr spectroscopy is a rather insensitive method, the simplicity of ^{13}C nmr makes this technique well suited for detecting minor concentration of isomeric species. Once *again* it is worth noting that ^{1}H-nmr is not suitable since the anomeric carbon centres are not protonated. Interestingly, the anomeric signals of the furanose forms in D-fructose appear at lower field than do those of the pyranose forms, a distinction that has proved valuable in the study of different sugars.[52] The dependence of ^{13}C-shifts on the orientation of the hydroxyl substituents (see Appendix A) also leads to characteristic differences in shieldings of anomeric centres in α-versus β-isomers. The ^{13}C chemical shifts in these molecules, which stem, in part, from non-bonded interactions, are, from one point of view, a measure of

H

ground state energetics. This philosophy has led to an interesting correlation between the *sum* of all the occurring chemical shifts in any one sugar and conformational free energies, with the largest value of the sum corresponding to the largest free energy.[52] This result is explained by the assumption that introduction of destabilizing interactions in the pyranose ring induces increased shielding of the ^{13}C nuclei around the ring. A knowledge of the ^{13}C characteristics of the lower molecular-weight sugars and their derivatives has led to a greater understanding of the composition and sequencing of several glucans.[52b] In principle, this is the same approach (monomer → intermediate → polymer) that has been used in the study of polypeptides.

High-resolution experiments for synthetic high polymers have been of special value in the determination of steric configurations in homopolymers and of sequence distributions in copolymers.[53] Schaefer[54] has applied ^{13}C nmr in the former problem on high conversion commercial probes of polyacrylo nitrile. In the ^{13}C-spectra of the nitrile carbons three lines of

LVI

LVII

LVIII

relative intensity 3 : 5 : 2 could be detected and assigned to the three sterically possible configurations (in the order of increasing field) isotactic, heterotactic and syndiotactic triads (LVI, LVII and LVIII).

We once again return to the inherent advantage of spectral simplicity offered by the ^{13}C{^{1}H} experiment. Even in cases where complete interpretation may not be possible a ^{13}C "fingerprint" of such acrylic polymers can provide a basis for detecting subtle structural changes in complicated molecules.

6.3 Nitrogen-15

Although, by far, the less abundant of the two nitrogen isotopes ($I = \frac{1}{2}$,

TABLE 6.2

^{15}N Chemical Shifts in the Triazenes, p-X-Ph-$N_1 = N_2 - N_3$ Me$_2$

X	N_1	N_2	N_3
NO$_2$	320	428	138·0
Cl	327·3	423·6	128·9
H	332·1	424·1	127·3
OCH$_3$		421·1	124·4

natural abundance = 0·36%) ^{15}N studies, in both enriched materials and in natural abundance, have been steadily increasing. In principle this nucleus offers advantages similar to those observed for ^{13}C. Specifically, it has a wide range of chemical shifts ($\sim$900 ppm), narrow line widths and a marked sensitivity to subtle changes either in molecular structure or electron density at nitrogen.

A simple example of a ^{15}N chemical shift/charge density correlation concerns the triazenes, $p - X - Ph - N_1 = N_2 - N_3$ Me$_2$, whose molecules are known to demonstrate restricted rotation around the $N_2 - N_3$ bond. While it is necessary to measure at various temperatures to clearly observe this hindered rotation in the ^{1}H spectrum, the room temperature ^{15}N data clearly show[55] the differing participations of all three nitrogen atoms in the extended π-system as a function of X (see Table 6.2).

The resonances of N_2 and N_3 shift downfield with increasing electron withdrawing ability of X, whereas the signal for N_1 shifts upfield. While for N_3 this is crudely in keeping with expectations for contributions such as LIX, the observed changes in δN_1 are important since they suggest that a structure such as LX may also be involved.

LIX

Advantage was taken of the favourable spin of ^{15}N to measure some one-bond coupling constants to platinum-195 ($I = \frac{1}{2}$, natural abundance = 33·7%). These studies were prompted by earlier phosphorus nmr studies on bis phosphine complexes of platinum which demonstrated a strong dependence of the values $^1J(^{195}$Pt, ^{31}P) upon the geometry of the complex.

$$\underline{LX}$$

The point at issue is whether the differences in the coupling constants should be attributed solely to a $d\pi - d\pi$ back bonding argument (e.g. LXIII).

$$\nu_{Pt,P} \sim 2.2{-}2.4\,KHz \qquad\qquad \nu_{Pt,P} \sim 3.4{-}3.6\,KHz$$

$$\underline{LXI} \qquad\qquad\qquad \underline{LXII}$$

In this ideology the two *trans* groups (PR_3 v. Cl^- in the *cis* isomer, PR_3 v. PR_3 in the *trans* case) compete for the $d\pi$ density on the metal, with a con-comitant increase in $^1J(Pt, P)$ for the winner.† Alternately, the observed differences may be accounted for based solely upon a rehybridization of the metal σ-bonding orbitals. This implies that the composition of the MO

$$\underline{LXIII}$$

used by the metal to bond to phosphorus in LXI differs from that used in LXII. Clearly, if such a π-bonding argument is crucial to the observation of significant differences in the donor atom-platinum coupling constants then one might expect little or no change in this type of coupling when the two ligands are not expected to be involved in π-bonding.

The ^{15}N spectra of the corresponding complexes of dodecylamine-^{15}N, $[PtX_2(C_{12}H_{27}{}^{15}N)_2]$, $X = Cl^-$, Br^- showed $^1J(^{195}Pt, {}^{15}N)$ values for the *cis* complexes of the order of 20% greater than for the *trans* analogs (e.g. 350 Hz v. 290 Hz) demonstrating that π-bonding is not a necessary condition for affecting changes in such coupling constants.[57]

If one assumes that $^{15}NO_2^-$ complexes give representative $^1J(Pt, {}^{15}N)$ values for sp^2 nitrogen bound to platinum (*trans*-$Pt(^{15}NO_2)_2(PBu_3)_2 = 453$ Hz, $Pt(^{15}NO_2)_4{}^{2-} = 592$ Hz) the magnitude of the nitrogen–platinum

† It is reasoned that more density on phosphorus via back bonding should increase donation from phosphorus to platinum via the phosphorus lone pair, thus strengthening the σ-component and increasing $^1J(Pt, P)$.

LXIV   LXV

coupling can be used to assign structure. This approach was used in an effort to distinguish between LXIV and LXV, the two isomeric platinum complexes derived from the reaction of $PtCl_4^{2-}$ with unsymmetrical o, o'-dihydroxy-azobenzenes. In practice it is frequently possible to synthesize[60] both isomers, that is with the metal bound either to N_1 or N_2. While the isomers may be quite different (colour, chromatography characteristics), the decision as to which isomer has N_1 free and which N_1 coordinated is not readily made. Synthesis of the materials enriched at N_1 with ^{15}N, followed by the measurement of the ^{15}N spectra readily provided the answer.[61] One spectrum revealed a nitrogen resonance flanked by two platinum-195 satellites separated by 523 Hz (LXIV) whereas in the second isomer only the main band was visible ($^2J(^{15}N, Pt)$ too small to be resolved in this situation).

A second rather "tricky" structural problem involving complexed molecular nitrogen was handled nicely using ^{15}N nmr.[60a] It was recognized from ^1H and ^{13}C studies that cooling a toluene-d_8 solution of the permethyltitanocene complex (LXVI) from $-52°$ to $-62°$ splits both the CH_3 proton and CH_3 ^{13}C resonances into two absorptions of approximately equal intensity. That one of these is indeed the "end-on" nitrogen adduct (LXVIa) was demonstrated from the ^{15}N (enriched) spectrum which showed the expected two doublets at 84·6 and 134·0 ppm resulting from coupling between the two non-equivalent nitrogen atoms. In addition the spectrum revealed a sharp singlet, thought to be the ^{15}N resonance of the "edge-on"

LXVI a   LXVI b

complex (LXVIb), although this latter assignment has been criticized.[60b] Clearly ^{15}N nmr is both a useful structural and electron density probe for nitrogen complexes of platinum.

$$\left[\begin{array}{l} \text{D-Phe—L-Pro—L-Val—L-Orn—L-Leu} \\ \text{L-Leu—L-Orn—L-Val—L-Pro—D-Phe} \end{array}\right]$$

<u>LXVII</u>

Occasionally, the appearance and just the appearance of a nitrogen spectrum, relative to the ^{13}C or ^{1}H analog, may suggest experiments which lead to useful chemical information. In the course of some fundamental ^{15}N studies,[61] it was observed that the natural abundance ^{1}H decoupled ^{15}N-spectrum of the cyclic peptide Gramacidin-S (LXVII) in methanol, showed ^{15}N resonances whose NOE's were of the order of 3·9 (maximum), thus suggesting a dominance of the dipole–dipole relaxation mechanism and the fulfilment of the extreme narrowing condition ($\omega\tau_c \ll 1$). However, in DMSO solution the "backbone" ^{15}N resonances of this peptide were no longer visible. The spectra are shown in Fig. 6.8. This nulling of signals for ^{15}N (which has a negative gyromagnetic ratio, γ) is possible if the NOE has changed to $+1$ as this will cause cancellation of the normally negative signal. This was shown to be the case using the gated decoupling technique, described in Chapter 3, which allows the measurement of a decoupled spectrum while suppressing the Overhauser "enhancement". The implication from the experiment in DMSO is that the correlation time, τ_c, has increased and that the value $\omega\tau_c$ is no longer relatively small. Additional ^{13}C measurements in combination with previous calculations permitted these workers to calculate τ_c values of $\sim 6 \times 10^{-10}$ and $5\cdot4 \times 10^{-9}$ s for this molecule in methanol and DMSO, respectively, thereby gaining insight into the effects of solvent on molecular motions. Similar variations in the intensity of ^{15}N signals due to changes in NOE have been observed for some hydrazine derivatives.[55]

6.4 Phosphorus-31

While phosphorus-31 nmr studies have always been moderately abundant ($I = \frac{1}{2}$, 100% abundant), FT nmr in conjunction with ^{1}H decoupling techniques, has brought this nucleus to the point where spectra on 1 mg of material (MW = 500–1000) are feasible. The structural potential of this nmr tool is easily demonstrated. Figure 6.9 shows the ^{31}P nmr spectrum of an intact muscle from a rat leg at different intervals after the excision.[55a] Most obvious is the marked decrease in the intensity of line III which is thought to correspond to creatine phosphate. Clearly, such *in vivo* measurements have much to offer the biochemist seeking information on the activity of

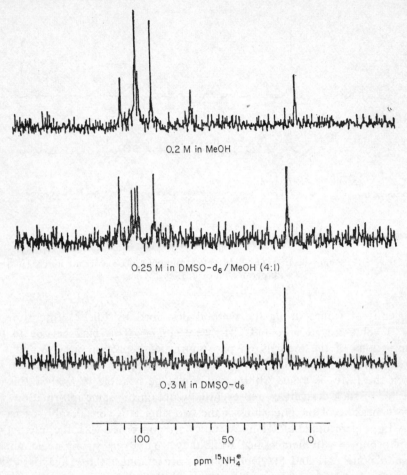

0.2 M in MeOH

0.25 M in DMSO-d_6 / MeOH (4:1)

0.3 M in DMSO-d_6

ppm $^{15}NH_4^{\oplus}$

FIG. 6.8. Natural abundance $^{15}N\{^1H\}$ spectra of gramicidin-S.[61]

phosphate controlled enzymes. Equally obvious is the necessity of measuring relatively quickly since the sample is literally "dying" before our eyes.

A somewhat more fundamental and certainly less dramatic question concerns the conformational free energies of cyclic systems containing phosphorus functions such as LXVIII. Thus at 183°K, when $X = H$, only the conformer having PR_2 equatorially disposed could be observed.[62] However, for the *cis* isomer when $X = CH_3$, $R = H$, CH_3, Cl the low temperature ^1H decoupled ^{31}P spectrum revealed two sharp signals whose relative intensities could be used to calculate $K_{equil.}$ and thus ΔG^0. These values were found to be in good agreement with those obtained by measuring the *cis* and *trans* 4-*t*-butyl derivatives and recalculating $K_{equil.}$, and sub-

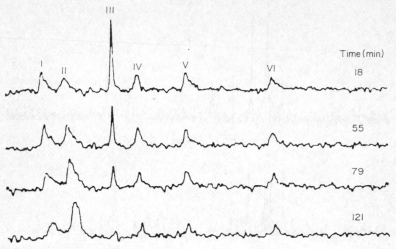

FIG. 6.9. FT ^{31}P nmr spectrum at 129 MHz of an intact muscle from the hind leg of a rat. (With permission of *J. Coord. Chem.*)

sequently *A* values, using the method described by Eliel.[63] Interestingly, the *A* values for *R* = H and CH$_3$ were similar. This may be due to the interactions of the methyls on phosphorus with protons at C-2 and C-3 when PMe$_2$ is equatorial, as well as the classical 1,3 and 1,5 interactions when the PMe$_2$ is axial. Although it might be possible to analyse the ^{1}H spectra of such derivatives and eventually obtain the same information, the direct measure of the intensities of the two singlets is considerably easier.

A large percentage of ^{31}P nmr studies are concerned with the interaction of phosphines with metals since, while they are not very strong nucleophiles, monodentate aryl and alkyl phosphines are amongst those ligands which bind most strongly to transition metals. Although the characterization of such complexes can be exceedingly difficult using conventional spectroscopic and ^{1}H nrm techniques, the relative simplicity of the ^{31}P H spectrum considerably reduces the complexity of the problem. The range of ^{31}P chemical shifts in such complexes is large (200 ppm) as is the sensitivity of δ^{31}P to changes in molecular structure. Additionally, the one bond metal–phosphorous coupling, when observable, is a useful probe for both molecular structure and complex geometry. (^{107}Ag, ^{109}Ag, ^{103}Rh, ^{183}W, ^{195}Pt and ^{199}Hg all have $I = \frac{1}{2}$ and couple with ^{31}P.)

In the chemistry of transition metal complexes of polydentate phosphines, ^1H decoupled ^{31}P nmr has provided both gross and subtle structural information from a single measurement. Thus, in the reaction of the tetradentate ligand QP (LXIX) with $W(CO)_6$, the primary product gave the ^{31}P spectrum shown in Fig. 6.10.[64] Somewhat surprisingly the spectrum reveals resonances for four distinct types of phosphorus, three of which are flanked by ^{183}W ($I = \frac{1}{2}$, natural abundance $\sim 14\%$) satellites. Extraction of all of the ^{31}P chemical shifts and phosphorus–tungsten couplings allows the following conclusions.

QP ≡

LXIX   LXX   LXXI

1. Since three of the four different types of phosphorus appear significantly downfield relative to the free ligand (a characteristic of phosphorus bound to tungsten), with the fourth resonating at the position of uncomplexed aromatic phosphine (14·4 ppm to *high* field of H_3PO_4), this potentially tetradentate ligand is acting as a tri-dentate. Furthermore, since the central phosphorus (70·6 ppm) has assumed its normal position at considerably lower field (20–30 ppm in these systems) than the complexed peripheral phosphorus atoms, it is one of these latter ligand atoms which is *not* complexed.

2. Of the two possible isomers, LXX and LXXI, the facial one (LXX) is most likely correct since all three of the observed $^1J(^{183}W, {}^{31}P)$ values (217–228 Hz) have magnitudes consistent with phosphorus *trans* to carbonyl (220–230 Hz) and not with phosphorus *trans* to phosphorus (260–270 Hz).[65]

3. The non-coordinated phosphorus, P_3, is not symmetrically disposed between P_2 and P_{21} (see LXX), and therefore these two phosphine groupings are not equivalent.†

Since the report by Wilkinson[66] almost ten years ago of the rapid hydrogenation of olefins by the homogeneous catalyst $RhCl(PPh_3)_3$, a considerable effort has been made to understand more fully how this catalyst functions. In this context ^{31}P nmr studies have been able to make an important contribution. The ^{31}P spectrum of the aforementioned square planar Rh(I) complex (LXXII) in solution shows a pair of doublets assignable to two

† This non equivalence is most probably related to the bulkiness of the ligand.

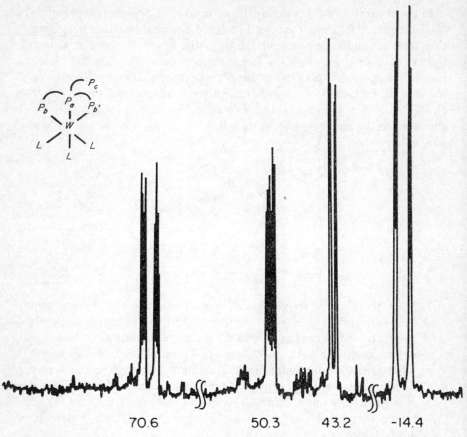

FIG. 6.10. The ^{31}P nmr spectrum of $[W(CO)_3 QP]$.[64]   Small resonances symmetrically disposed about the main bands are ^{183}W satellites. (With permission of *Nature*.)

equivalent ^{31}P nuclei and a pair of triplets due to the unique phosphorus consistent with structure **LXXII**. Each type of phosphorus reveals a large doublet which results from the one bond rhodium-103 ($I = \frac{1}{2}$) interaction, in addition to the AX_2 sub-spectrum which stems from the phosphorus–phosphorus coupling. Although small quantities of free ligand, and thus some phosphine exchange, could be detected, there was no evidence for significant amounts of a species such as $RhCl(PPh_3)_2$ as might have first been thought.[66] Under 100 psig of H_2 Wilkinson's catalyst (LXXII) has been shown[67] to react to give a complex which can be unambiguously assigned from its $^{31}P\{^1H\}$ and ^1H spectra. The product (LXIII) revealed an $A_2 BX$ type ^{31}P spectrum with a reduced value of $^1J(Rh, P)$ as well as two different types of hydride proton (τ 19·8 and 27·3) in the ^1H spectrum, one of which shows the expected relatively large coupling $^2J(P, H)$ value (152 Hz) associated with phosphorus *trans* to hydride. The facile reaction of LXXII to LXXIII and the failure to observe any change in the ^{31}P spectrum of LXXII in the presence or absence of cyclohexene under the conditions of homogeneous hydrogenation are suggested[67] to be evidence in favour of the "hydride route" mechanism. This description involves the initial reaction of LXXII with H_2 prior to olefin coordination.

LXXIV

In a similar vein ^{31}P and ^1H studies have shown[67] that the addition of H_2 to the symmetrical dimer $[RhCl(P(p-Tol)_3)_2]_2$ gives LXXIV, $L = P(p-tol)_3$, and not a tetrahydride as first thought.

The extensive occurrence of intramolecular *ortho*-metallation reactions involving activation and cleavage of *ortho*-carbon–hydrogen bonds in aromatic nitrogen and phosphine complexes of transition metals is frequently readily detected using ^{31}P nmr. Thus both palladium and platinum complexes of the type *cis*-$[MX_2(P(OAr)_3)_2]$ react in boiling decalin with the elimination of HX to give structures of type LXXV. The observation of two *cis* coupled

LXXV

phosphite resonances ($^2J(P, P) \sim$ 30–60 Hz, *trans* $^2J(P, P) >$ 100 Hz) and, for the case where M = Pt, the magnitude of the values $^1J(Pt, P)$ strongly indicate the nature and geometry of the complex. The detection of *two* types of phosphorus, a trivial conclusion after seeing the ^{31}P spectrum leads quickly to the correct structure. Before leaving this type of chemistry it is worthwhile pointing out that the ^{13}C chemical shift and intensity† of the metallated aromatic carbon signal may provide strong evidence that such a

† Probably smaller since carbons without directly bound protons frequently have longer spin lattice relaxation times.

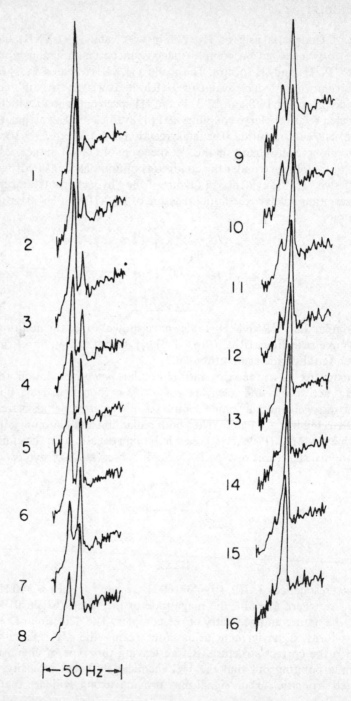

FIG. 6.11. Stopped flow Fourier transform nmr spectra obtained by mixing 78 mM PABE with 20 mg/ml α-chymotrypsin (final pH = 6·9). Spectral width shown in 50 Hz. The total time between each spectrum is 4·86 seconds.[69]

reaction has occurred. This is especially true in cases where the integration of the ^{1}H spectrum is not trivial.

6.5 " Other " nuclei

While nmr is frequently used in the study of time dependent chemical phenomena, the time scale of the reaction must be relatively long to permit the observation of several (e.g. reactants, intermediates *and* products) species in solution. The FT nmr technique is more suitable than the CW method since a relatively wide band of frequencies is simultaneously irradiated and the data collected in a short time period. Thus it has been shown[69] that ^{1}H FT nmr techniques can be made compatible with "stopped flow" measurements such that reactions with half-lives as short as 1 s may be monitored. A specific example involves the hydrolysis of L-phenylalanine-*t*-butyl ester (PABE) by α-chymotrypsin. The essential nmr spectra are shown in Fig. 6.11. In this experiment 78 mM L-phenylalanine-*t*-butyl ester was mixed with 20 mg/ml α-chymotrypsin (final pH = 6.9) to give the spectra shown. In these experiments one is monitoring the time dependence of both the disappearance of ester and the emergence of one of the products, *tertiary*-butanol. At such concentrations perhaps the most suitable (and most sensitive) nucleus is ^{1}H although it may be anticipated that the instrumentation necessary for the experiment might be refined in the future to allow the facile study of other nuclei using stopped flow methods.

The number of nuclei capable of study using FT techniques has recently been extended to include, amongst others, metals such as ^{199}Hg,[70] ^{207}Pb,[71] ^{113}Cd,[72] ^{205}Tl[73] and ^{195}Pt.[74] Such studies are interesting because the chemical shift range of metals is of the order of *thousands* of ppm with substituent effect/chemical shift correlations of the order of hundreds of ppm. Although much of this work is in the preliminary stage, early efforts have suggested that a significant chemical return for time invested may be forthcoming. Some examples for one of these metals, ^{195}Pt, are illustrative. The ^{195}Pt chemical shifts of XLVIII and XLIX are almost identical suggesting[42] that the metal–olefin bond strengths are quite similar. This conclusion is in agreement with some, but contrasts with other nmr studies on these molecules and highlights the necessity of a multinuclear approach. ^{195}Pt nmr has also been used[74] to confirm that the solvation of both Zeise's salt and Zeise's dimer with acetone yields the same end product, most probably LXXVI, via, in the latter case, solvent cleavage of the halogen bridge and in the former, solvation and subsequent replacement of Cl$^-$ by acetone.†

In the absence of an organic "handle", ^{195}Pt nmr can provide an interesting alternative to uv spectroscopy.

The aquo specie, Na[PtCl$_3$(OH$_2$)], which results from solvolysis of Na$_2$PtCl$_4$

$$Na_2PtCl_4 \underset{-H_2O}{\overset{H_2O}{\rightleftharpoons}} Na[PtCl_3(OH_2)] + NaCl$$

is readily observed[74] in the ^{195}Pt spectrum of the latter complex.

In the absence of added Cl$^-$ ion, the aquo specie may have the same abundance as the parent tetrachloroplatinate. Thus simple inspection of ^{195}Pt nmr spectrum can provide data with regard to potential intermediates in solution. Similar studies on other metals will undoubtedly reap the same type of harvest.

References

1. J. B. Stothers, "Carbon-13 NMR Spectroscopy." Academic Press, New York and London, 1972.
2. G. C. Levy and G. L. Nelson, "Carbon-13 Nuclear Magnetic Resonance for Organic Chemists." Wiley Interscience, New York, 1972.
3. E. Breitmaier and W. Voelter, "^{13}C NMR Spectroscopy." Verlag Chemie, Weinheim, 1974.
4. J. T. Clerc, E. Pretsch and S. Sternhell, "^{13}C-Kernresonanzspektroskopie." Akad. Verlagsgesellschaft, Frankfurt, 1973.
5. "Topics in Carbon-13 NMR Spectroscopy" (Ed., G. C. Levy). Wiley Interscience, New York, 1974, Vol. 1.
6. J. N. Shoolery, Varian Application Note, NMR-75-3.
7. J. N. Shoolery, Varian Application Note, NMR-73-5.
8. J. Ehrenfreund, M. P. Zink and H. R. Wolf, *Helv. Chim. Acta*, **57**, 1048 (1975).
9. G. Maier, "Valenzisomerisierungen." Verlag Chemie, Weinheim, 1972.
10. H. Guenther, H. Schmickler, W. Bremser, F. A. Straube and E. Vogel, *Angew, Chem.*, **85**, 585 (1973); *Angew. Chem. Internat. Edit.*, **12**, 570 (1973).
11. H. Guenther, B. D. Tunsgal, M. Regitz, H. Scherer and T. Keller, *Angew. Chem.*, **83**, 585 (1971); *Angew. Chem. Internat. Edit.*, **10**, 563 (1971).
12. K. Grohmann, J. B. Gruetzner and J. D. Roberts, *Tetrahedron Lett.*, 1969, 917.
13. J. F. M. Oth, *Pure Appl. Chem.*, **25**, 573 (1971).
14. E. Vogel, H. Koenigshofen, K. Muellen and J. F. M. Oth, *Angew. Chem.*, **86**, 229 (1974); *Angew. Chem. Internat. Edit.*, **13**, 281 (1974).
15. E. Vogel, J. Wassen, H. Koenigshofen and K. Muellen, J. F. M. Oth, *Angew. Chem.*, **86**, 777 (1974); *Angew. Chem. Internat. Edit.*, **13**, 732 (1974).

† Not necessarily in that order.

16. H. Gerlach and K. Muellen, *Helv. Chim. Acta*, **57**, 2234 (1974).
17. D. A. House and J. W. Blunt, *Inorg. Nucl. Chem. Letters*, **11**, 219 (1974).
18. K. S. Dhami and J. B. Stothers, *Can. J. Chem.*, **43**, 479 (1965); K. S. Dhami and J. B. Stothers, *Can. J. Chem.*, **43**, 498 (1965).
19. D. H. Marr and J. B. Stothers, *Can. J. Chem.*, **43**, 596 (1965).
20. H. Guenther, Physikalische Methoden in der Chemie: Kohlenstoff-13-NMR II, in *Chemie in unserer Zeit*, **8**, (3), 84 (1974).
21. E. A. Williams, J. D. Cargioli and A. Ewo, *J.C.S. Chem. Commun.* 1975, 366.
22. E. Heilbronner and H. Bock, "Das HMO-Modell und seine Anwendung." Verlag Chemie, Weinheim, 1968.
23. H. Guenther, Lecture held at the 2nd IUPAC-Conference on non-benzenoid aromatic compounds, Lindau, Germany, 1974.
24. A. J. Jones, T. D. Alger, D. M. Grant and W. M. Litchman, *J. Amer. Chem. Soc.*, **92**, 2386 (1970); A. J. Jones, P. D. Gardner, D. M. Grant, W. M. Litchman and V. Boeckelheide, *J. Amer. Chem. Soc.*, **92**, 2395 (1970).
25. T. D. Alger, D. M. Grant and E. G. Paul, *J. Amer. Chem. Soc.*, **88**, 5397 (1966).
26. F. J. Weigert and J. D. Roberts, *J. Amer. Chem. Soc.*, **92**, 1347 (1970); J. B. Grutzner, M. Jautelat, J. B. Dence, R. A. Smith and J. D. Roberts, *J. Amer. Chem. Soc.*, **92**, 7107 (1970).
27. G. B. Savitsky, K. Namikawa and G. Zweifel, *J. Phys. Chem.*, **69**, 3105 (1965).
28. S. Berger and A. Rieker, *Tetrahedron*, **28**, 3123 (1972).
29. T. Kametani, M. Kajiwara, T. Takahashi and K. Fukomoto, *Tetrahedron*, **31**, 949 (1975).
30. G. A. Olah and A. M. White, *J. Amer. Chem. Soc.*, **90**, 1884 (1968); G. A. Olah and A. M. White, ibid., **91**, 5801 (1969); G. A. Olah and R. D. Porter, ibid., **93**, 6877 (1971); R. Ditchfield and D. P. Miller, ibid., **93**, 5287 (1971).
31. G. A. Olah and P. W. Westerman, *J. Amer. Chem. Soc.*, **95**, 7530 (1973); G. A. Olah, P. W. Westerman and J. Nishimura, ibid., **96**, 3548 (1974).
32. J. F. Wolf, P. G. Harch, R. W. Taft and W. J. Hehre, *J. Amer. Chem. Soc.*, **97**, 2902 (1975).
33. H. C. Brown and E. N. Peters, *J. Amer. Chem. Soc.*, **95**, 2400 (1973).
34. G. A. Olah and G. D. Mateescu, *J. Amer. Chem. Soc.*, **92**, 1430 (1970).
35. J. F. M. Oth, K. Muellen, H. Koenigshofen, J. Wassen and E. Vogel, *Helv. Chim. Acta*, **57**, 2387 (1974).
36. R. A. Goodman, E. Oldfield and A. Allerhand, *J. Amer. Chem. Soc.*, **95**, 7553 (1973).
37. D. Dodrell and W. S. Caughey, *J. Amer. Chem. Soc.*, **94**, 2510 (1972); see also: W. S. Caughey, J. O. Alben, W. Y. Fujimoto and J. L. York, *J. Org. Chem.*, **31**, 2631 (1966).
38. S. G. Boxer, G. L. Closs and J. J. Katz, *J. Amer. Chem. Soc.*, **96**, 7058 (1974).
39. J. A. Connor, E. M. Jones, E. W. Randall and E. Rosenberg, *J.C.S. Dalton Trans.* 1972, 2419.
40. M. H. Chisholm, H. C. Clark, J. E. H. Ward and K. Yasufuku, *Inorg. Chem.* **14**, 893 (1975); see also: B. E. Mann, in *Advances in Org. Metl. Chem.*, **12**, 135 (1974); M. H. Chisholm, H. C. Clark, L. E. Manzer, J. B. Stothers and J. E. H. Ward, *J. Amer. Chem. Soc.*, **95**, 8574 (1973).
41. D. G. Cooper, G. K. Hamer, J. Powell and W. F. Reynolds, *J.C.S. Chem. Commun.* 1973, 449.
42. P. S. Pregosin and L. M. Venanzi, *Helv. Chim. Acta*, **58**, 1548 (1975).
43. P. S. Pregosin and E. W. Randall, "Nuclear Magnetic Resonance of Nuclei other than Protons" (Eds, T. Axenrod and G. A. Webb). Wiley Interscience, New York, 1974, p. 248.

44. D. K. Dalling and D. M. Grant, *J. Amer. Chem. Soc.*, **89,** 6612 (1967).
45. J. D. Roberts, F. J. Weigert, J. I. Kroschwitz and H. J. Reich, *J. Amer. Chem. Soc.*, **92,** 1338 (1970).
46. G. E. Maciel, *in* "Topics in Carbon-13 NMR Spectroscopy" (Ed., G. C. Levy). Wiley Interscience, New York, 1974. Chapter 2.
47. R. Deslauriers and I. C. P. Smith, "Topics in Carbon-13 NMR Spectroscopy" (Ed., G. C. Levy). Wiley-Interscience, New York, 1974, Chapter 1 and references therein.
48. K. Wuethrich, *Pure Appl. Chem.*, **37,** 235 (1974).
49. C. C. McDonald and W. D. Phillips, *J. Amer. Chem. Soc.*, **91,** 1513 (1969).
50. H. S. Isbell and W. W. Pigman, *J. Res. Nat. Bur. Stand*, **20,** 773 (1938).
51. D. Dodrell and A. Allerhand, *J. Amer. Chem. Soc.*, **93,** 2779 (1971).
52a. J. F. Stoddart, "Stereochemistry of Carbohydrates." Wiley Interscience. New York, 1971, Chapter 4.
52b. H. J. Jennings and I. C. P. Smith, *J. Amer. Chem. Soc.*, **95,** 606 (1973); ibid., **96,** 8081 (1975).
53. J. Schaefer, The carbon-13 NMR analysis of synthetic high polymers, *in* "Topics in Carbon-13 NMR Spectroscopy" (Ed., G. C. Levy). Wiley Interscience. New York, 1974, Vol. 1.
54. J. Schaefer, *Macromolecules*, **4,** 105 (1971).
55a. D. I. Hoult, S. J. W. Busby, D. G. Gaddian, G. K. Radda, R. E. Richards and P. J. Seely, *Nature*, **252,** 285 (1974).
55b. T. Axenrod and P. S. Pregosin, unpublished results.
56. A. Pidock, R. E. Richards and L. M. Venanzi, *J. Chem. Soc.*, 1707 (1966).
57. P. S. Pregosin, H. Omura and L. M. Venanzi, *J. Amer. Chem. Soc.*, **95,** 2047 (1973).
58. E. Steiner and G. Schetty, submitted for publ., *Helv. Chim. Acta.*
59. E. Steiner and P. S. Pregosin, *Helv. Chim. Acta.*, **59,** 376 (1976).
60a. J. E. Bercow, E. Rosenberg and J. D. Roberts, *J. Amer. Chem. Soc.*, **96,** 612 (1974).
60b. J. Mason and J. G. Vinter, *J.C.S. Dalton,* 2522 (1975).
61. G. E. Hawkes, W. M. Litchman and E. W. Randall, *J. Magn. Res.*, **19,** 255 (1975).
62. M. D. Gordon and L. D. Quinn, *J.C.S. Chem. Commun.* 1975, 35.
63. E. L. Eliel, *Chem. and Ind.* 1959, 568.
64. R. J. Mynott, P. S. Pregosin and L. M. Venanzi, *J. Coord. Chem.*, **3,** 145 (1973).
65. S. D. Grim and D. A. Wheatland, *Inorg. Chem.*, **8,** 1716 (1969).
66. J. A. Osborn, F. H. Jardine, J. F. Young and G. Wilkinson, *J. Chem. Soc. A,* 1711 (1966).
67. C. A. Tolman, P. Z. Meakin, D. L. Lindner and J. P. Jesson, *J. Amer. Chem. Soc.*, **96,** 2762 (1974).
68. N. Ahmad, E. W. Aincough, T. A. James and S. D. Robinson, *J. Chem. Soc. Dalton,* 1151 (1973).
69. J. J. Grimaldi and B. D. Sykes, *J. Amer. Chem. Soc.*, **97,** 273 (1973).
70. G. E. Maciel and M. Borzo, *J. Magn. Res.*, **10,** 388 (1973).
71. R. M. Hawk and R. R. Sharp, *J. Magn. Res.*, **10,** 384 (1975).
72. G. E. Maciel and M. Borzo, *J.C.S. Chem. Common.* 1973, 394.
73. L. W. Reeves, unpublished results.
74. W. Freeman, P. S. Pregosin, S. N. Sye and L. M. Venanzi, *J. Magn. Res.*, **22,** 473 (1976).

7

Dynamics and reaction mechanisms

The chemical shift characteristics of ^{13}C and ^{31}P nmr expand the potential for the investigation of reaction mechanisms since nonequivalent nuclei in simple and relatively complex systems can be detected as well-resolved lines. Thus the fate of a specific centre in the course of a complex biological reaction can be followed by isotopic labelling while systems undergoing exchange, whose ^{1}H nmr characteristics are unfavourable succumb more readily to lineshape analysis using the carbon nmr technique. We shall demonstrate the usefulness of proton decoupled ^{13}C- and ^{31}P-spectra in both these fields.

7.1 Rate processes

If intramolecular mobility gives rise to an equilibrium of two isomers, M and N, the signal from an individual nucleus, being once in M and once in N, can only be recorded separately under the following conditions.[1]

1. The nucleus must possess a different magnetic environment in M than N.

2. The difference in the chemical shifts $v_M - v_N$ must be large relative to the linewidth of the signals.

3. The rate of exchange must be small in comparison with $v_M - v_N$.

An illustrative interpretation of the situation can be provided by considering the lifetime τ of an individual nucleus in a given magnetic site. We will restrict ourselves to the simple case in which the nuclei are not J coupled and the magnetic sites are equally populated. From the lifetimes, τ_M and τ_N, of the nuclei in M and N we can define an average lifetime τ by

$$\tau = \frac{\tau_M \cdot \tau_N}{\tau_M + \tau_N}.$$

When the exchange process obeys the three conditions cited above this

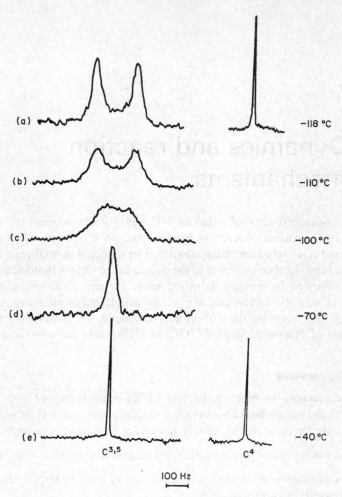

FIG. 7.1. Temperature dependence of the ^{13}C nmr C-3, C-5 and C-4 signals of pyrazole.[4]

situation is known as the "slow exchange limit" and implies that $\tau \gg 1/(v_M - v_N)$. If the rate of exchange, k, increases, τ decreases ($k = 1/\tau$) and the lines broaden and approach each other. When the signals have coalesced to afford a broad average resonance the lifetime is

$$\tau = \frac{\sqrt{2}}{2\pi} (v_M - v_N).$$

With increasing exchange rate the signal continues to narrow until, in the "fast exchange limit", a sharp signal appears. This implies

$$\tau \ll 1/(v_M - v_N)$$

and the resonance frequency is now the average value of v_M and v_N. This process can clearly be followed in Fig. 7.1. From lineshape analysis of the signals in the exchange region we can find the individual exchange rates, since the equilibrium of isomers M and N, according to absolute rate theory, depends on the temperature and on the enthalpy of activation of the iso-merization process, respectively.[2, 3] If the temperature dependence of exchange rates is then determined the thermodynamic parameters of inter-conversion may be derived. For a detailed review of dynamic nmr, the reader should consult the dedicated literature[1] since the view described above is quite simple. From the features of dynamic nmr that have been outlined here the difference in resonance frequencies $v_M - v_N$ is obviously an essential para-meter. Since the large ratio of spectral width to line width for non-proton nuclear resonances facilitates separation of signals in different isomers, ^{13}C, ^{15}N, ^{31}P ... nmr have a distinct advantage over ^1H nmr. Moreover, dynamic ^1H nmr suffers from the existence of spin–spin couplings. These very often not only render more difficult the qualitative detection of an exchange, but may also make the lineshape calculations impossible. In the case of proton decoupled ^{13}C or ^{31}P spectra, however, a specific magnetic site is represented by a single resonance.

An illustration is afforded by the proton transfer in pyrazoles[4] (LXXVII) which occurs so rapidly that, according to the ^1H nmr spectra, the molecules

LXXVIIa   LXXVIIb

exhibit an effective C_{2v} symmetry over a large temperature range; however, ^{13}C nmr conveniently allows us to follow this tautomeric process (see Fig. 7.1).

The kinetic parameters calculated from these ^{13}C data have permitted a detailed mechanistic description of the underlying proton-exchange, which process takes place via intermolecular self association of the solute.

As has been mentioned in Chapter 6, cyclic $4n - \pi$ systems, such as the perimeter compound 1,7-methano[12]annulene, possess localized π bonds,[5] however, even at relatively low temperatures a π-bond shift occurs which interconverts the two isodynamic structures LXXVIIIa and LXXVIIIb. This

$$\underline{LXXVIII}a \rightleftharpoons \underline{LXXVIII}b$$

causes exchange of the magnetic sites 2 and 6 as well as 3 and 5, respectively, whereas the magnetic sites of positions 1 and 4 are unaffected. This latter point is advantageous in that the signals of C-1 and C-4 can then provide an approximate internal measure of the natural linewidth of the system. Determination of the enthalpy of activation[5] of the double-bond migration in this system is of exceptional utility in interpreting the electronic structure of such molecules. This energy contains the "resonance destabilization" of the transition state which, according to a plausible mechanistic model, can be regarded as a delocalized $4n$-π-electron system.[1d, 5] Again we point out that it is exceedingly difficult to describe this process properly by means of ^{1}H nmr.

Problems of molecular conformation occasionally introduce complications if more than one rate process is included. This is what is envisaged in the stereo-dynamics of directly attached nitrogen atoms in the bicyclic biurethanes

$$\underline{LXXIX}$$

(LXXIX).[6] At elevated temperatures C-5 and C-6 give rise to a sharp singlet which, at about 8°C, splits into two peaks of equal intensity. On further cooling the lower field signal splits into two new absorptions. The process at lower temperature is due to the hindered amide $(>N-CO)$ rotation, while the high temperature process is interpreted as a torsion around the $N-N$ bond and/or an out-of-plane distortion at two nitrogen atoms (essentially a double inversion).

The correct analysis of dynamic processes requires a knowledge of the exchange diagrams of the magnetic sites. This becomes quite complicated if the sites are differently populated or if they show different exchange rates. If classical example of this problem concerns bullvalene which is known to undergo a degenerate Cope-rearrangement.[7] Essentially, all three cyclopropane-bonds can be involved in this process (see Fig. 7.2). Due to the

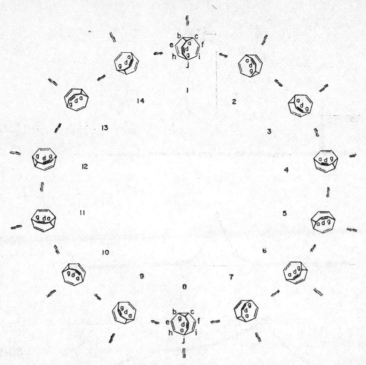

FIG. 7.2. A sequence of 14 isomerizations of bullvalene which restores the labelled atoms to their initial positions. (With permission of *Helv. Chim. Acta.*)

three possible isomerization pathways, any of the ten CH groups in the molecule can occupy any position in the course of time. As a consequence of this movement all CH groups become equivalent and at higher temperatures the ^{13}C-spectrum consists of a single sharp line (Fig. 7.3).[8]

The exchange diagram for the isomerization process in bullvalene (Fig. 7.4) can be constructed by considering one specifically labelled structure and the three directly connected ones. The arrows (one arrow/nucleus) in the diagram indicate how the different nuclei (identified by letters) exchange between the different magnetic sites (identified by numbers). The fact that any magnetic site in the ^{13}C spectrum is represented by a singlet becomes especially important if one considers the first line broadening in the slow exchange limit (Fig. 7.5). Not only can the isomerization rate be determined, but, since the broadening of a line representing magnetic site "i" is also directly proportional to the diagonal element a_{ii} of the stochastic exchange matrix, we can, as well, obtain direct information concerning the exchange mechanism.[8]

Variable temperature ^{13}C-nmr has proved an especially valuable technique

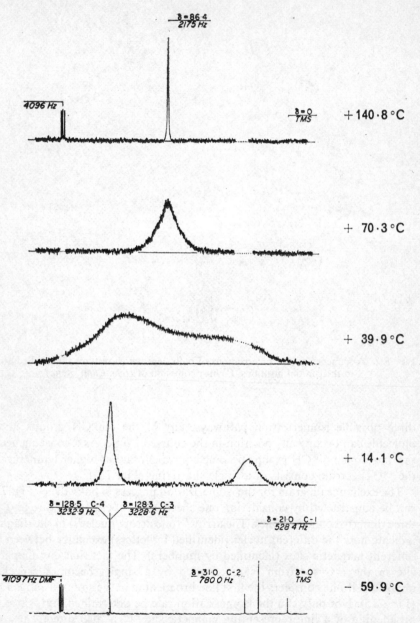

FIG. 7.3. The $^{13}C\{^1H\}$ spectrum of bullvalene at different temperatures.[12] (With permission of *Helv. Chim. Acta.*)

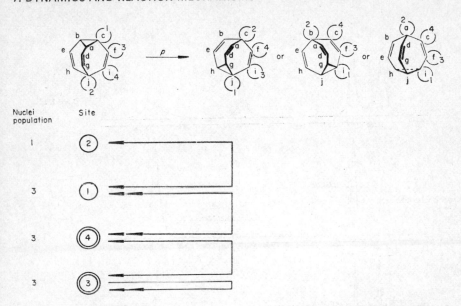

FIG. 7.4. The exchange diagram for the Cope rearrangement of bullvalene. (With permission of *Helv. Chim. Acta.*)

in the study of the dynamics of stereochemically non-rigid metal carbonyl complexes. An illustrative example of this concerns the observation of localized CO exchange at the iron and ruthenium centres of the complex $[H_2FeRu_3(CO)_{13}]$ (LXXX) (see Fig. 7.6).[9] At $-72°C$ the ^{13}C-nmr spectrum exhibits broadening of the iron carbonyl resonances (1), (2) and (3), whereas the signals of carbonyls located at ruthenium remain sharp. At $-45°C$ the signals of the latter coalesce and at 95°, we observed a single sharp signal representing fast exchange at all sites. The explanation offered for the three different exchange processes involves, first, a low temperature localized scrambling of carbonyl groups on Fe, probably due to an opening of the bridges on the ruthenium side, followed by a higher temperature CO-scrambling at Ru and finally, ligand exchange between both metals.

If ligands of transition metal complexes contain phosphorus, direct information concerning stereochemically non-rigid behaviour can be obtained by inspection of the temperature-dependent ^1H-decoupled ^{31}P-spectra. In complexes of the type $[(diene) M(CO)_2(EPTB)]$ where $(M = Fe$ or $Ru)$ diene = cyclohexadiene, cycloheptadiene and EPTB = LXXXI, the ^{31}P-spectra at ambient temperatures consist of a single sharp resonance.[10] As the temperature is lowered the signal broadens and finally splits into two peaks of unequal intensity. These peaks are thought to be due to phosphorus-atoms

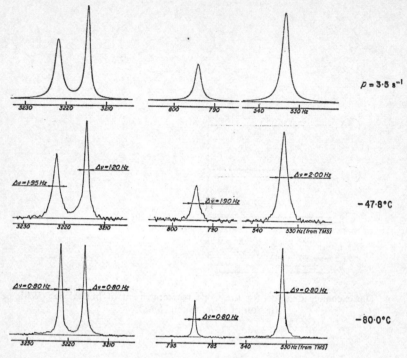

FIG. 7.5. The initial broadening of the individual ^{13}C resonances of bullvalene. (With permission of *Helv. Chim. Acta.*)

in the two isomeric forms, LXXXII and LXXXIII, of an idealized tetragonal-pyramidal coordination geometry.

The determination of the structure of complexes of the type ML_5, in solution, has been the subject of several inquiries.[11] The two most commonly accepted possibilities are the trigonal bipyramid (LXXXIV) and square

LXXX

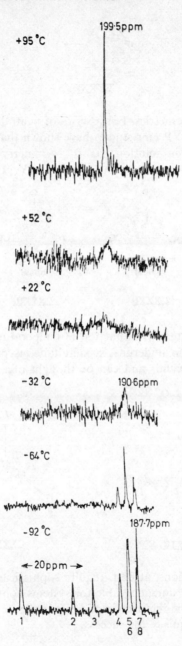

FIG. 7.6. Variable temperature ^{13}C nmr spectra of LXXXI from $-90°$C to $+95°$C in CHFCl$_2$—CD$_2$Cl$_2$ (50 mg in 3 cm^3) approximately 30% enriched ^{13}CO and 0·05 M in Cr(acac)$_3$.[9]

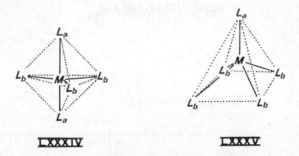

LXXXI

pyramid (LXXXV) geometries. For a series of neutral and cationic pentakis phosphite complexes, ^{31}P nmr studies have shown that, at low temperature, the limiting structure in solution has D_{3h} symmetry and gives rise to an $A_2 B_3$ type ^{31}P spectrum, consistent with LXXIV. The change in the line

apical

basal

LXXXII

LXXXIII

shape of the $A_2 B_3$ pattern as a function of temperature has been used,[11] to suggest that the system undergoes a simultaneous exchange of two axial with two equatorial ligands and can be thought of as undergoing a Berry rearrangement.

LXXXIV

LXXXV

All of these calculations, at best, require sophisticated computer analysis and frequently become unmanageable for systems containing many coupled spins. In this respect the relative simplicity provided by ^{13}C or ^{31}P spectroscopy may be a prerequisite for success.

7.2 Reaction intermediates

In addition to characterizing the structure of reaction intermediates in

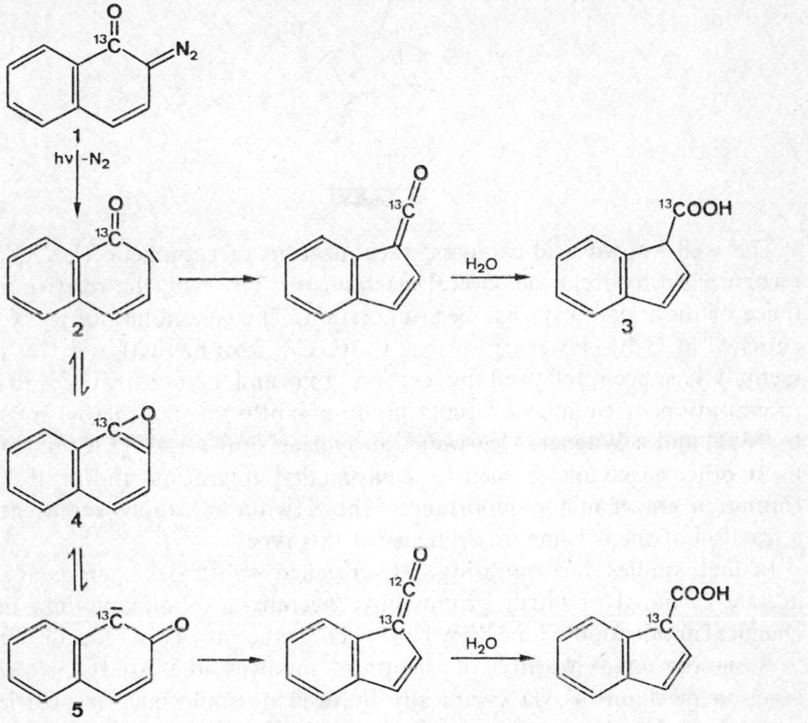

FIG. 7.7. The Wolff-rearrangement of an α-diazo-ketone as followed by a ^{13}C label.[12]

solution (e.g. carbocations), ^{13}C nmr provides a convenient method for tracing the pathway of reaction mechanisms via the use of enriched labels. An enriched carbon is readily detected and the ability to state at exactly which position the label has been incorporated can sometimes define the course of the reaction. The carbon label can be especially convenient in that no degradation is necessary for its detection and its initial insertion in biological systems can occasionally be done "naturally" by a microbe.

An illustrative example concerns the photolysis of 2-diazo 1-^{13}C naphthalene 1(2H)-one[12] (1 in Fig. 7.7). The potentially antiaromatic oxirenes (4) have been discussed as intermediates in the Wolff-rearrangement of α-diazoketones. Photolysis of aqueous solutions of (1) gives rise to a 1-indenecarboxylic acid which contains all the label in the carboxy group. This finding excludes the isomerization of the originally formed oxycarbene (2) to the oxycarbene (5). Either (2) does not undergo ring closure to the oxirane (4) or (4) reopens exclusively to (2).

LXXXVI

The well known acid-catalysed racemization of camphene (LXXXVI) is recognized[13] to occur via several mechanisms; however, the relative importance of these pathways has been uncertain. The scrambling of a ^{13}C label enriched at C-10 between positions C-10, C-8 (*exo*-methyl) and C-9 (*endo*-methyl) has been followed by carbon nmr and demonstrates[13] that the racemization of camphene occurs predominantly via *exo*-methyl migration (~53%) and a Wagner–Meerwein 2,6-hydride shift (~41%). Contributions from other mechanisms such as *endo*-methyl migration and/or tricyclene formation are of minor importance. These two applications represent only a fraction of the existing experiments of this type.

In fact studies incorporating ^{13}C-enriched substrates such as formate, acetate carbonyl or nitrile groups have become so abundant that, in biochemical applications alone, they have become the subject of recent reviews.[14]

A newer area worthy of mention involves the study of radical reaction mechanisms via chemically induced dynamic nuclear polarization methods.[15] This has been especially useful in cases involving reversible freeradical reactions where reagent and product are identical. While there are some difficulties[16] it is reasonable to assume the FT nmr will be increasingly employed in this area since in the longer period of time necessary for a CW spectrum some of the nuclear polarization is lost. In addition, the measurement of the ^{13}C resonance of a fully substituted carbon provides important mechanistic information when the radical attack occurs at this centre. Such a ^{13}C study has already been shown to be useful.[17]

We conclude that, at many levels, the multinuclear approach is a fruitful one and that at the root of such measurements lies the FT nmr technique.

References

1a. C. S. Johnson Jr., *Advan. Magn. Res.*, **1**, 33 (1965).
1b. G. Binsch, *in* "Topics in Stereochemistry" (Eds., E. L. Eliel and N. L. Allinger). Wiley Interscience, New York, 1968.
1c. J. F. M. Oth, *Pure Appl. Chem.*, **25**, 573 (1971).
2. F. A. L. Anet and R. Anet, Configuration and conformation by nmr, *in* "Determination of Organic Structures by Physical Methods" (Eds., F. C. Nachod and J. J. Zuckerman). Academic Press, New York and London, 1971, Vol. 3.
3. H. Kessler, *Angew. Chem.*, **82**, 237 (1970); *Agnew. Chem. Internat. Edit.*, **9**, 219 (1970).

4. A. N. Nesmeyanov, E. B. Yavelovich, V. N. Babin, N. S. Kochetkova and E. J. Fedin, *Tetrahedron*, **31**, 1461 (1975).

5. E. Vogel, H. Koenigshofen, K. Muellen and J. F. M. Oth, *Angew. Chem.*, **86**, 229 (1974); *Angew. Chem. Internat. Edit.*, **13**, 281 (1974).

6. Y. Nomura, N. Masai and Y. Takeuchi, *J.C.S. Chem. Commun.*, 1975, 307.

7. W. E. Doering and W. R. Roth, *Angew. Chem.*, **75**, 27 (1963); *Tetrahedron*, **19**, (1963); G. Schroeder, *Chem. Ber.*, **97**, 3140 (1964); R. Merenyi, J. F. M. Oth and G. Schroeder, *Chem. Ber.*, **97**, 3150 (1964).

8. J. F. M. Oth, K. Mullen, J. M. Gilles and G. Schroeder, *Helv. Chim. Acta*, **57**, 1415 (1974).

9. L. Milone, S. Aime, E. W. Randall and E. Rosenberg, *J.C.S. Chem. Commun.* 1975, 452.

10. T. H. Whitesides and R. A. Budnik, *Inorg. Chem.*, **14**, 64 (1975).

11. P. Mcakin, A. D. English, S. D. Ittel and J. P. Jesson, *J. Amer. Chem. Soc.*, **97**, 1254 (1975); P. Meakin and J. P. Jesson, *J. Amer. Chem. Soc.*, **96**, 5751 (1974); J. P. Jesson, ibid., **96**, 5760 (1974).

12. K. P. Zeller, *J.C.S. Chem. Commun.* 1975, 317.

13. C. W. David, B. W. Everling, R. J. Kilian, J. B. Stothers and W. R. Vaughan, *J. Amer. Chem. Soc.*, **95**, 1265 (1973).

14. M. Tanabe, H. Seto and L. R. Johnson, *J. Amer. Chem. Soc.*, **92**, 2157 (1970); among recent papers see for example: K. G. R. Pachler, P. S. Steyn, R. Vleggaar and P. L. Wessels, *J.C.S. Chem. Commun.* 1975, 355; M. R. Adams and J. D. Bu'Look, ibid, 1975, 389; J. S. E. Holker and K. Young, ibid. 1975, 525.

15. R. Kaptein, "Advances in Free Radical Chemistry" (Ed., G. H. Williams). Elek., London, Vol. 5, p. 319.

16. R. R. Ernst, W. P. Aue, E. Bartholdi, A. Hoehener and S. Schacublin, *Pure Appl. Chem.*, **37**, 47 (1974).

17. R. Kaptein, R. Freeman, H. D. W. Hill and J. Bargon, *J.C.S. Chem. Commun.* 1973, 953.

Appendix A

Comments on chemical shifts of nuclei other than ^1H

In discussing the applications of studies involving ^{13}C, ^{15}N, ^{31}P and other nuclei we have avoided detailed analysis of any one spectrum since the individual screening characteristics for a given nucleus[1] and a given class of compounds vary considerably. There are, however, some terms which are common to all these nuclei and, although this area is quite independent of FT nmr, we feel that it is useful to review these points.

The frequency at which resonance occurs for a given nucleus has been shown in Chapter 1 to be:

$$v = \gamma H_0 (1 - \sigma_t)/2\pi \qquad (A1.1)$$

where σ_t is the total screening constant. This screening term has several components (see A1.2)

$$\sigma_t = \sigma_d + \sigma_p + \sigma_x \qquad (A1.2)$$

the most important of which, for ^1H, are σ_d, the diamagnetic screening term and σ_x, a catch-all term in which are included local anisotropic effects, etc. For most other nuclei the term σ_p (see A1.3) is far and away the most important component of σ_t. This quantity is generally formulated as

$$\sigma_p \propto - \frac{1}{\Delta E} \langle r^{-3} \rangle \sum_B Q_{AB} \qquad (A1.3)$$

and contains expressions involving an "average" triplet excitation energy, ΔE, the average value of the inverse cube of the non-s-orbital radius, $\langle 1/r^3 \rangle$ and the elements of the atomic orbital charge density bond order matrix, Q_{AB}.

For molecules which possess a relatively low lying excited state (e.g. $n \rightarrow \pi^*$, $\pi \rightarrow \pi^*$), the ΔE term may make a significant contribution to σ_p and consequently decrease the observed shielding. Thus the ^{13}C resonance of carbonyl compounds appear at lower field than do those for olefin derivatives which, in turn are deshielded relative to aliphatic carbons. Similarly, the nitrogen resonance of nitrosobenzene, PhNO, falls at much lower field than the corresponding nitrogen absorption of nitrobenzene which in its turn is significantly lower than aniline.[1]

The effect of charge and the terms Q_{AB} is perhaps best illustrated by comparing the carbon resonance position of the carbonium ions, $R_3 C^+$, with any other carbon position. Before such cationic species were measured the "normal" range of ^{13}C chemical shifts was thought to be of the order of 250 ppm with most values falling between 0 and 200 ppm. The measurement of the central carbon of $(CH_3)_3 C^+$ at ~340 ppm $[(CH_3)_3 CH \sim 25 \cdot 2 \text{ ppm}]$ has widened this range considerably. Carbon-13 chemical shift dependences in the range 160–200 ppm/electron have been suggested.

While less dramatic in magnitude the effects of alkyl substituents on the ^{13}C chemical shift have proved extremely important. These effects are commonly denoted as the α, β and γ substituent effects[3] and referred initially to the effect, at each of these positions, of substituting a methyl group for a hydrogen atom at the α position (see a). Subsequently, groups other than CH_3 have been considered. For a number of nuclei ($X = \,^{13}$C, ^{15}N, ^{31}P (β and γ only)) it has been found that the α and β effects are *deshielding* while the γ effect is *shielding* in nature.

(a)

(b)   (c)

A number of papers[1] have recorded the magnitude of these effects with the ^{13}C substituent effects[3] receiving the most attention. It seems generally agreed upon that the γ effect has its source in the steric compressions of proximate C–H bonds. Whatever the source, these substituent effects (at the moment only for ^{13}C but most probably for other nuclei as well) have been shown to be dependent upon both conformation and configuration and thus have become most useful structural probes (see Chapter 6). The ^{13}C chemical shift of the γ carbon in cyclohexyl systems such as, b, can be 5–6 ppm to higher field for $Y = CH_3$, OH, NH_2 than in structure c. For a more thorough understanding of the changes that these substituents induce in σ, the reader is recommended to consult reference 3.

Quite naturally the dominance of one term does not exclude other contributions to σ_t. Indeed, it has been suggested that in order to properly appreciate the significance of the σ_t term a σ_d correction should be included.[4]

In the interpretation of *small* differences in chemical shift (of the order of several ppm) we would do well not to ignore contributions from local anisotropic and medium effects. Where the nuclei under consideration are transition metals, these latter "effects" may be of a moderate size since solvents and counter ions may sometimes complex with the metal (e.g. the solvent effect on the ^{195}Pt chemical shift in the $PtCl_4^{2-}$ is of the order of tens of ppm).[5]

Appendix B

Signal assignment in carbon nmr

It is clear from the examples given in Chapter 6 that ^{13}C spectra normally exhibit many more lines than do ^{31}P or ^{15}N spectra, so that, relatively speaking, the question of signal assignment becomes more important for this nucleus.

The assignment is normally performed using a combination of several techniques amongst which the following double resonance experiments are frequently applied: (a) off-resonance coherent decoupling, (b) selective decoupling and (c) recording of completely undecoupled spectra. These methods are considered in Chapter 3. It is clear, that the differentiation of

$$-CH_3, \quad \diagdown CH_2, \quad -CH \text{ or } -C-$$

fragments by inspection of peak multiplicities in the off-resonance decoupled ^{13}C spectra works, in principle,† analogously for

$$-NH_2, \quad \diagdown NH \text{ or } -N-$$

groups in the nitrogen spectrum.

The identification of quaternary carbons is an important problem in the spectra of complex organic or bio-organic molecules, and can be achieved simply by applying decreased power to the broad band decoupler. Since less decoupling power is necessary to decouple these resonances they are selectively enhanced. The same result is sometimes obtainable by "chemical" suppression of long-range ^{13}C-^{1}H-couplings with paramagnetic reagents (see Chapter 5).

The "X" nucleus chemical shift is a highly indicative parameter for describing molecular structure and the detection of its signal in a "typical" shift region, as with protons, is characteristic. For carbon-13 the first major distinction is one of hybridization, with the sp^3, sp and sp^2 carbons found at

† Proton exchange can complicate matters.

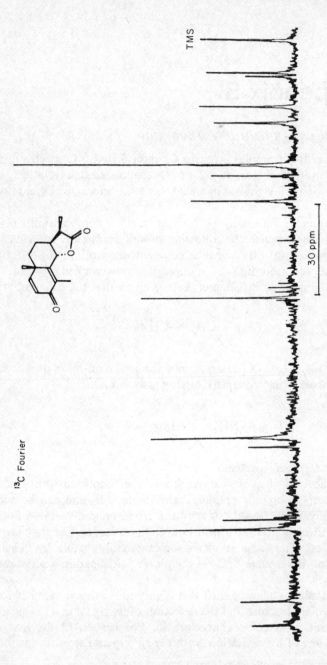

FIG. B.1. The $^{13}C\{^1H\}$ spectrum of β-santonin. (With permission of Wiley Interscience.)

successively lower field. Typical ranges are 0–80, 90–110 and 110–220 ppm, respectively. Within these groupings come the effects of charge density and electronegativity. The former usually induces chemical shift changes, *for the atom bearing the charge*, in the expected sense (e.g. more positive charge yields lower field shifts). Care must be taken in considering the effects of charge on adjacent carbon centres as the direction of the change is frequently inverted (e.g. RCO_2H at *higher* field than RCO_2^- but RCH_2NH_2 at *lower* field than $RCH_2NH_3^+$). The electronegativity of the substituent exerts a classical effect and this is typified by the examples for halogen (F > Cl > Br) and oxygen v. nitrogen (0 > N, ~ 40 ppm to low field v. ~ 30 ppm). Somewhat smaller, but quite important are the alkyl substituent effects described in Appendix A.

Still quite common in assigning ^{13}C spectra of higher molecular weight compounds is the use of model systems whereby one assumes that a carbon in a particular fragment of a small molecule has essentially the same chemical shift in the larger analog if no additional chemical and/or physical interactions are present.

All of these techniques were used in the assignment of the ^{13}C spectrum (see Fig. B.1) of the mold metabolite β santonin: $C_{15}H_{18}O_3$, d. The proton decoupled ^{13}C spectrum of this material shows 14 sharp singlets one of which resulted from the overlap of two resonances. Six of these resonances are below 100 ppm with the remaining nine above 80 ppm. These two groups may be assigned to the sp^2 and sp^3 carbons, respectively.

(d)

Within the former group there are two signals at very low field (> 170 ppm) and four at higher field which correspond to the carbonyl and olefin carbons, respectively. An off-resonance spectrum revealed that only two (C−1, C−2) of the six sp^2 carbons had directly bound protons. In addition, it was observed that the four olefinic carbons were observed as two groups of two each (151·9, 155·0 ppm and 126·0, 128·8 ppm). We recall, from Chapter 6, that in α, β unsaturated ketones the β carbon possesses some positive charge due to a resonance structure such as XXIII, and therefore carbons 1 and 5 might be expected to, and indeed *do* appear at lower field than carbons 2 and 4. Therefore, based on off resonance (C−1, C−2 from C−4, C−5) and expected charge densities (C−1, C−5 from C−2, C−4), the olefinic carbons are assigned. Since it is known that ketones always appear 20–40 ppm to lower field than esters (perhaps in part due to stabilization of some positive charge on carbon by the second oxygen) the assignment of the sp^2 carbons is complete.

Turning to the aliphatic region one can immediately separate the quaternary carbon, C−10 from the groups C−6, C−7, C−11 (methine); C−8, C−9 (methylene) and C−13, C−14, C−15 (methyl) from the off-resonance spectrum measured previously. Carbon 6 is the lowest field aliphatic resonance stemming from its connection to the strongly electronegative oxygen. The *proton* at C−11 has a lower field shift than that proton located at C−7 and thus *selective* (cw) decoupling experiments allowed an assignment of these two carbons (in a similar fashion one can pick out C−15 from the other two methyl groups) and therefore completed the assignment of the methine carbons. Carbon 9 appears at lower field than C−8 due to the β methyl substituent effect of C−14 (see Appendix A) and so it remained only to distinguish the C−13 and C−14 methyl groups. This might be possible using substituent effects but such a calculation proved not to be necessary since the position of C−14 is characteristic for this type of molecular grouping. In the set of compounds α, β, α-epi and β-epi santonin the resonance position of the C−14 methyl group varies only slightly while that for C−13, which is relatively close to and involved in, major structural alteration in these molecules, changes significantly. With this last comparison (which also assists in establishing the identity of C−15) the assignment was completed.

In addition to these commonly used methods there are several special techniques which are becoming increasingly popular. One such method, although occasionally requiring extensive synthetic work, involves specific deuteration. In the $^{13}C\{^1H\}$ experiment the coupling $^1J(^{13}C\text{-D})$ is unaffected so that the carbon atom bearing the deuteron, when observable,† is readily assigned. In addition to this coupling, whose value is $\approx {}^1J(^{13}C\text{-H})/6\cdot5$, deuterium has an isotopic effect on the carbon chemical shift. The sense of this effect is upfield with a magnitude of $0\cdot1$–$1\cdot0$ ppm. The introduction of deuterium at selected sites has proved quite valuable in carbohydrate chemistry.[6]

A second method involves the use of lanthanide shift reagents and in some cases, aqueous metal ions themselves. The principles involved in the use of these probes are the same for all nuclei and will not be described here.

Lastly, we mention, again, that the spin lattice relaxation time, T_1, can be quite useful in assigning ^{13}C spectra and we refer the reader to Chapter 4 for details.

† In addition to longer T_1 values, signals from carbon nuclei bearing deuterons have weaker intensity due to the loss of NOE and the coupling to deuterium.

Appendix C

The relationship between τ_c and T_1 for spin-lattice relaxation mechanisms other than dipole–dipole

We give here the relationships which are analogous to that for the dipole–dipole term in Chapter 4 when other relaxation mechanisms are dominant. We will in all cases assume $\omega \tau_c \ll 1$.

1. *Spin rotation* (for liquids undergoing isotropic molecular reorientation)

$$1/T_1 = (2\pi I T/h^2)C_{\text{eff}}^2 \, \tau_J,$$

where I = moment of inertia of the molecule, C_{eff}^2 (spin rotation tensor) = $(2C_\perp^2 + C_\parallel^2)/3$, τ_J = angular momentum correlation time $(\tau_c \tau_J = I/6kT)$, T = absolute temperature.

2. *Chemical shift anisotropy*

$$1/T_1 = (2/15)\gamma^2 H_0^2 (\sigma_\parallel - \sigma_\perp)^2 \tau_c,$$

where $\sigma_\parallel$ and $\sigma_\perp$ are the parallel and perpendicular components of the screening tensor σ (assumed to be axially symmetric).

3. *Scalar coupling*

$$1/T_1 = (2A^2/3)S(S+1)(\tau_{\text{sc}}/[1 + (\omega_I - \omega_2)^2 \, \tau_{\text{sc}}^2]),$$

where I and S are the two coupled nuclei, A = the coupling constant $(2\pi J)$, τ_{sc} = the correlation time for scalar coupling (= chemical exchange time when an exchange modulates; = to the quadrupolar relaxation time when a nuclear quadrupole modulates.)

4. *Nuclear quadrupole*

$$1/T_1 = (3/40)\left(\frac{2I+3}{2I-1}\right)\left(1 + \frac{\eta^2}{3}\right)(1/I^2)\left(\frac{e^2 qQ}{\hbar}\right)^2 \tau_c,$$

where η = the assymetry parameter, $\left(\dfrac{e^2 qQ}{\hbar}\right)$ is the quadrupole coupling constant.

Appendix D

Comments on quadrature detection

We have already mentioned (Chapters 2 and 3) the disadvantages associated with the inability of the computer to differentiate between the frequencies (carrier wave $+ v$) and (carrier wave $- v$). A useful method to resolve this ambiguity of sign involves simultaneously recording the output of two phase detectors whose reference signals are in quadrature, i.e. $\cos \omega t$ and $- \sin \omega t$. The two outputs, which are 90° out of phase and can be considered as the real and imaginary components of the magnetization, are then fed through different digitizers before being stored separately. A variety of suggestions on how one should treat these data have been made and the reader is recommended to consult reference 7 for details. After complex Fourier transformation a spectrum without folded lines is obtained.

Using this detecting scheme the carrier wave is placed in the *centre* of the spectrum. This is advantageous in that we may sample more slowly than when the carrier is set on one side of the spectrum, for the same spectral band width. Additionally, we increase the likelihood that the pulse power is distributed uniformly across the spectrum (see Chapter 3). Last, and perhaps most important, we gain in signal-to-noise by $\sqrt{2}$ ($\approx 40\%$) since noise which would otherwise appear in the spectrum, via folding, has been eliminated. A second approach to the signal-to-noise problem involves the installation of a sharp (crystal) filter which passes only the desired bandwidth while excluding folded noise.

Although the quadrature detection technique is of a somewhat specialized nature, the likelihood that such detecting systems will be included in newer commercial spectrometers, makes it worthy of note.

References

1. (a) See "The Determination of Organic Structures by Physical Methods" (Eds, F. C. Nachod and J. J. Zuckerman). Academic Press, New York and London, 1971, Chapters 4, 6 and 7. See also (b) R. Ditchfield and P. Ellis, *in* "Topics in Carbon-13 NMR Spectroscopy" (Ed., G. C. Levy). Wiley Interscience, New York, 1974, Vol. 1, Chapter 1.
2. A. Saika and C. P. Slichter, *J. Chem. Phys.*, **22**, 26 (1954); M. Karplus and J. A. Pople, *J. Chem. Phys.*, **38**, 2803 (1963); J. A. Pople, *J. Chem. Phys.*, **37**, 53, 60 (1962); *Mol. Phys.*, **7**, 301 (1964); T. K. Wu, *J. Chem. Phys.*, **49**, 1139 (1968); **51**, 3622 (1969).
3. G. Maciel, *in* "Topics in Carbon-13 NMR Spectroscopy" (Ed., G. C. Levy). Wiley Interscience, New York, 1974, Vol. 1, Chapter 2.
4. R. Grinter and J. Mason, *J. Chem. Soc. A*, 2196 (1970); J. Mason, *ibid.*, 1038 (1971).
5. W. Freeman, P. S. Pregosin, S. N. Sze and L. M. Venanzi, *J. Magn. Res.*, **82**, 473 (1976).
6. H. J. Koch and A. S. Perlin, *Carbohydr. Res.*, **15**, 403 (1970); J. Lehmann, *Carbohydr. Res.*, **2**, 1 (1966); P. A. J. Gorin, *Can. J. Chem.*, **52**, 458 (1974); A. S. Perlin, B. Casu and H. J. Koch, *Can. J. Chem.*, **48**, 2596 (1970); D. E. Dorman and J. D. Roberts, *J. Amer. Chem. Soc.*, **92**, 1355 (1970).
7. A. G. Redfield and R. K. Gupta, *Adv. Magn. Res.*, **5**, 81 (1971); J. D. Ellett Jr., M. G. Gibby, U. Haeberen, L. M. Huber, M. Mehring, A. Pines and J. S. Waugh, *Adv. Magn. Res.*, **5**, 117 (1971); E. O. Stejskal and J. Schaefer, *J. Magn. Res.*, **14**, 160 (1974); A. G. Redfield and S. D. Künz, *J. Magn. Res.*, **19**, 250 (1975); D. I. Hoult, *J. Magn. Res.*, **21**, 337 (1976).

Answers to problems

1a. While there may be some loss of NOE, relative to the *ortho* and *meta* carbons, the smaller intensities of the substituted carbons stem from their longer T_1 values.

1b. A reduction in the pulse length and/or the introduction of a post delay will help to remedy the problem.

2a. "Pulse breakthrough" will frequently result in a rolling baseline. A short pre-delay, t_{pre}, of the order of 1–2 dwell times is often helpful.

3. Some form of data manipulation is called for; however, the cards seem stacked against us. A relatively large negative TC value (sensitivity enhancement) might broaden the already poorly resolved intense signals. Conversely, a relatively large positive value for TC might reduce the already weak signals to the level of the noise. We should, flexibility permitting, perform each operation in turn, thus finishing with "3" spectra. This problem should serve to remind us that a single spectroscopic result may arise from "n" measurements and $m \cdot n$ manipulations.

4. There are many acceptable answers. Higher resolution (perhaps measurement at a different magnetic field strength) careful integration and lanthanide shift reagents all might be useful. In practice we observed the ^{13}C spectrum under conditions of both ^1H and ^{31}P decoupling. The collapse of the two doublets gave a readily interpretable spectrum.

5. Since γ_{15N} is negative, the NOE observed will depend upon the extent to which the term $(\gamma_{1H}/\gamma_{15N})/2$ is either larger or smaller than $+1$ (see Chapter 2). If dipole-dipole (^{15}N–H) relaxation is important (implying that a W_2 process is relatively efficient) $(1 + [(\gamma_{1H}/\gamma_{15N})/2])$ can be as much as $-3\cdot93$. Thus the observed ^{15}N signals will be either positive or negative depending upon the T_1 relaxation pathways for the different nitrogen atoms. Chapter 4 deals with some consequences of dipole–dipole relaxation.

Subject Index

A

γ—*See* Gyromagnetic ratio of the nucleus

μ—*See* Nuclear magnetic moment

ω—*See* Resonance frequency

ω/γ—*See* Fictitious field

Acetic acid,
^{13}C T_1 value, 47
p-chlorophenylhydrazide, ^{13}C spectrum, 53
phenylhydrazide, 23

Acetone, ^{13}C T_1 value, 47

Acetone-d_6 as lock substrate, 29

Acquisition time, 17

Adenosine 5'-monophosphate, T_1 value, 49

Adiabatic passage experiment, 5

Aldehydes, ^{13}C chemical shifts and, 94

Alkanes, spin-lattice relaxation time, 70

Ammonium chloride-^{15}N as test substance, 49

Ammonium, decyltrimethyl-, ^{13}C T_1 value, 47

Analog-to-digital conversions, 15

Acetonitrile, ^{13}C T_1 value, 47

Acetophenone, ^{13}C T_1 value, 47

Acids (*see also* Acetic acid)
^{13}C chemical shifts and, 94

Angular momentum, 2

Aniline, ^{15}N resonance, 132

Aniline-^{15}N, p-chloro-, ^{13}C spectrum, 57

Annulenes, ^{13}C n.m.r., 91

Apodization routine, 25

Arsonium, tetramethyl-, bromide, ^{75}As spectrum, 58

Averaging (*see also* Signal averaging) 12

Azulene, ^{13}C n.m.r., 95

B

Basic pancreatic inhibitor, ^{13}C n.m.r., 102

Benzene
^{13}C chemical shifts, 96
^{13}C T_1 value, 48

Benzene-d_6 as lock substrate, 29, 49

Benzene, ethyl-
as test substance, 49
^{13}C spectrum, 32, 33
^{13}C T_1 value, 48
flip angle and, 46
n.m.r., 7

—, ethynyl-
^{13}C T_1 value, 48

—, nitro-
^{13}C T_1 value, 48
^{15}N resonance, 132

—, nitroso-
^{15}N resonance, 132

Benzocyclobutene, ^{13}C n.m.r., 96

Benzoyl chloride, ^{13}C T_1 value, 47

Benzylamine, N,N'-dimethyl-, n.m.r., 9

Berry rearrangement, 128

Biphenyl, 3-bromo-, spin-lattice relaxation time, 72

Biurethanes, bicyclic, ^{13}C n.m.r., 122

Bloch equation, 65

Bromoform, ^{13}C T_1 value, 47

Bullvalene
Cope-rearrangement, 125
^{13}C n.m.r., 124, 126
^{13}C{^{1}H} spectrum, temperature and, 124

Butane, 1-bromo-, ^{13}C T_1 value, 47

C

Cadmium, ^{113}Cd, n.m.r., 115
Camphene, racemization, ^{13}C n.m.r., 130
Carbenes, metal complexes, ^{13}C n.m.r., 98
Carbohydrates
  deuteration, ^{13}C n.m.r. and, 138
  ^{13}C n.m.r., 103
Carbon
  ^{13}C, chemical shifts, 132
  ^{13}C n.m.r., signal assignment in, 135–138
Carbonium ions
  ^{13}C n.m.r. and, 96
  ^{13}C resonance, 133
Carbon suboxide, ^{13}C n.m.r., 95
Carbonyl complexes
  iron ruthenium, CO exchange, ^{13}C n.m.r., 125
  metal, ^{13}C n.m.r., 125
Carbonyl compounds
  ^{13}C chemical shifts and, 94, 132
Carrier wave, selecting the position of, 41, 42
Chemical shifts, 2, 85, 132–134
  ^{13}C, 87
  reactivity and, 96
  ^{13}C, structural parameters and, 94–104
Chemical shift anisotropy, 62, 63, 139
Chloroform, ^{13}C T_1 value, 47
Chloroform-d as lock substrate, 29
Chloroform-^{13}C, inversion of energy level populations in, 39
Chlorophyll a, ^{13}C n.m.r., 98
Cholesteryl chloride
  spin-lattice relaxation time, 73
  T_1 value, 49
Chromium (III)-acetyl acetonate as paramagnetic reagent, 79, 80
α-Chymotrypsin, reaction with phenylalanine tert-butyl ester, 114, 115
Cobalt (II)-acetate as paramagnetic reagent, 79
Cobalt, bis(ethylenediamine)-
  ^{13}C n.m.r., 93
—, tris(glycinato)-
  ^{13}C n.m.r., 93

Computers, 15–26
Continuous wave experiment, 5
Cooley-Tukey algorithm, 25
Cope-rearrangement
  degenerate, bullvalene, 125
  bullvalene, ^{13}C n.m.r., 122–125
Correlation time, 64
Coupling constant, 2, 8, 19
Cyano groups, ^{13}C n.m.r., 90
Cycloheptatriene, ^{13}C n.m.r., 91
Cyclohexane
  ^{13}C T_1 value, 48
  methyl derivatives, ^{13}C n.m.r., 107
—, methoxy-
  ^{13}C T_1 value, 48
Cyclohexanol
  ^{13}C T_1 value, 48
  methyl derivatives, ^{13}C n.m.r., 107
Cyclo-octane, methoxy-, ^{13}C T_1 value, 48
Cyclo-octanol, ^{13}C T_1 value, 48
Cyclo-octatetraene
  dianion, ^{13}C chemical shifts, 96
  dimer, ^{13}C n.m.r., 91
Cyclopentadienyl anion, ^{13}C chemical shifts, 96
Cyclopentane, methoxy-, ^{13}C T_1 value, 48
Cyclopentanol
  ^{13}C T_1 value, 48
  spin-lattice relaxation time, 70
Cyclopropyl cation, ^{13}C n.m.r., 97

D

Data handling, inversion recovery and, 67
Data input, 15–17
Data manipulation, 19–26, 55
Decane
  ^{13}C spectrum, 84
  ^{13}C T_1 value, 47
  ^{1}H spectrum, 84
—, 1-bromo-
  ^{13}C T_1 value, 47

G

E

H

F